Cooling Technologies - Technologies and Systems to Guarantee Thermal Comfort in Efficient Buildings

Edited by David Bienvenido-Huertas

Published in London, United Kingdom

Cooling Technologies - Technologies and Systems to Guarantee Thermal Comfort in Efficient Buildings
http://dx.doi.org/10.5772/intechopen.105254
Edited by David Bienvenido-Huertas

Contributors
Ali F. Ali Fadiel, Yoonis A.M. Esham, Giuseppe Starace, Silvia Macchitella, Gianpiero Colangelo, Mikhail Luchko, Raksha Manel Shenoy, Julio César Rincón-Martínez, Armando Núñez-de Anda, Francisco Fernández-Melchor, Paula Scherer, Daiana de Oliveira Fauro, Giane de Campos Grigoletti

First published in London, United Kingdom, 2023 by IntechOpen
IntechOpen is the global imprint of INTECHOPEN LIMITED, registered in England and Wales, registration number: 11086078, 5 Princes Gate Court, London, SW7 2QJ, United Kingdom

British Library Cataloguing-in-Publication Data
A catalogue record for this book is available from the British Library

Additional hard and PDF copies can be obtained from orders@intechopen.com

Cooling Technologies - Technologies and Systems to Guarantee Thermal Comfort in Efficient Buildings
Edited by David Bienvenido-Huertas
p. cm.
Print ISBN 978-1-83769-583-6
Online ISBN 978-1-83769-582-9
eBook (PDF) ISBN 978-1-83769-584-3

Meet the editor

David Bienvenido-Huertas is an assistant professor in the Department of Building Construction, University of Granada, Spain. His areas of expertise include climate change in the building sector, adaptive thermal comfort, heat transfer, fuel poverty, energy conservation measures, and the design of nearly zero-energy buildings. He is an author of more than fifty research papers and a recognized reviewer of various internationally indexed journals.

Contents

Preface

Buildings have a great environmental impact due to their high energy consumption. Much of the consumption is due to refrigeration systems. Rising outdoor temperatures, extended summers, and heat waves are increasing the demand for cooling energy. This is leading to a greater need to use refrigeration systems. To achieve a low-carbon society, it is necessary to improve refrigeration systems. Reducing compressor consumption, using refrigerants with lower Global Warming Potential (GWP), and reducing distribution losses are some of the challenges that refrigeration systems face. As such, many researchers are focusing their studies on the analysis and optimization of cooling technologies.

This book is a compendium of existing research on cooling systems and their implications for the objectives of society in the 21st century. It presents research, studies, reviews, and case studies related to refrigeration systems from the perspectives of design and technology and energy efficiency and thermal comfort.

David Bienvenido-Huertas
Assistant Professor,
Department of Building Construction,
University of Granada,
Granada, Spain

Section 1

Design and Technology

Chapter 1

Absorption Refrigeration Cycle Technology

Ali F. Ali Fadiel and Yoonis A.M. Esham

Abstract

Absorption refrigeration technology was introduced to address serious issues such as the energy crisis, rising fuel prices, and environmental concerns associated with traditional compression refrigeration systems. They are attracting increasing attention due to advantages such as the use of low-mass heat sources and environmentally friendly working materials. However, this technology has two major drawbacks, including the very large size of the refrigeration unit and a low coefficient of performance (COP), which hinder the commercial success of the absorption system. Considerable research has been done to develop strategies to improve the COP of absorption systems in order to make absorption refrigeration technology more competitive than conventional compression refrigeration systems. This chapter aims to provide a literature review on whether this technique is an effective and promising solution for augmentation.

Keywords: absorption refrigeration, coefficient of performance, water/LiBr, single-effect machines, coefficient of performance (COP)

1. Introduction

Absorption cooling research has attracted the attention of many investigators in recent years as a result of the great growth of producing an environmentally friendly cooling system. The severity of the ozone depletion problem caused by CFCs and HCFCs has prompted the rapid development of CFC-free air conditioning technology. In terms of energy use, preventing global warming requires an overhaul of energy use practices to increase efficiency. From this point of view, absorption refrigeration research has attracted the attention of many researchers because it does not use ozone-depleting substances as refrigerants and can minimize the use of fossil fuel electricity. Vapor absorption cooling system uses heat rather than mechanical work as the energy input. The working fluid is the refrigerant while the material with which it is in solution is absorbent. The absorbent can be either in the liquid or the solid phase. The refrigerant is generally in the liquid or vapor phase, depending upon the process it is undergoing in the refrigeration cycle.

The most widely used absorption chiller is the ammonia-water machine, which uses ammonia (Li-Br) as the refrigerant and water (H2O) as the distribution (absorption) medium. Other absorption refrigeration structures consist of water-lithium bromide and water-lithium chloride structures, where water serves as the coolant.

The latter systems are limited to air conditioning applications where the minimum temperature is above the freezing point of water.

The benefits of absorption systems have remained constant and include the following:

- Absorption machines have lower electrical needs compared to vapor compression systems.
- The input energy to absorption systems, is a low-grade form of energy (heat) and is readily available as waste in large power plants and factories. It is also available from alternative energy sources solar, biomass, geothermal, etc.
- Absorption units pose no global environmental ozone depletion and may have less impact on global warming than most other options.
- Absorption machines eliminate the concerns about lubricants in refrigerants.
- Absorption units are economically attractive because fuel costs are substantially less than electric costs. Absorption equipment, which is heat driven, can both cool and heat for human comfort and process control. Heat sources for driving an absorption unit include the following:
    - Direct gas and oil firing
    - Indirect steam heating from boilers
    - Steam turbine exhaust
    - Gas turbine exhaust
    - Waste process steam
    - Cogeneration heat recovery steam or hot water
    - Hot processes fluids
    - Heat recovery from process steams and flue gases.

Of course, a small amount of mechanical work is added to the absorption system in the form of energy required to drive the solution recirculation pump. However, this is far less than the work required to run the compressor in the compression system, since the liquid has a smaller volume than the equivalent mass of vapor.

With larger absorption units, water is usually cooled by passing it through the tubes of the evaporator. This cooled water then flows through air coils to air condition spaces fluid or through other exchangers to cool process fluids [1, 2].

2. Compression refrigeration cycles

The fully reversible Carnot cycle is an ideal model of a refrigeration cycle operating between two constant temperatures or between two fluids at different temperatures,

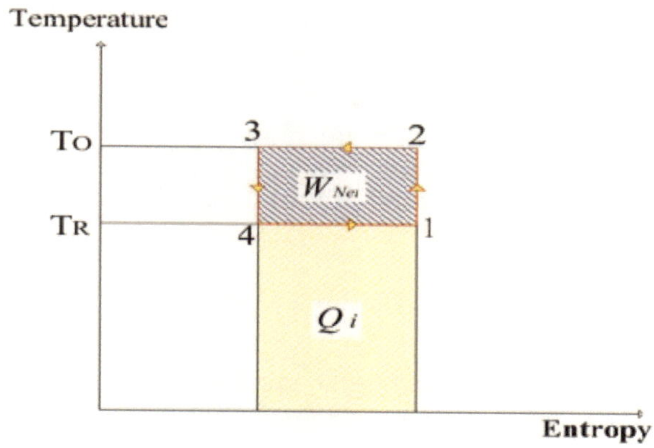

Figure 1.
Carnot cycle on temperature-entropy coordinates. Source: https://www.academia.edu/26602849/ASHRAE_HVAC_2001_Fundamentals_Handbook_pdf.

each having infinite heat capacity. Reversible cycles have two important properties: first, no refrigeration cycle can have a higher coefficient of performance than a reversible Carnot cycle operating within the same temperature limits, and second, all reversible cycles operating within the same temperature limits have the same coefficient of performance (**Figure 1**).

The heat is removed from the cooled area at a constant temperature T_R. At a constant ambient temperature T_O heat is emitted. The cycle ends with isentropic expansion and isentropic compression.

The output heat energy can be written as:

$$Q_O = T_O(S_2 - S_3) \tag{1}$$

The input heat energy can be written as:

$$Q_i = T_R(S_1 - S_4) = T_R(S_2 - S_3) \tag{2}$$

So, the network is equal to the difference between the heat energy output and input. And can be written as:

$$W_{net} = Q_o = Q_I \tag{3}$$

Therefore, the coefficient of performance can be written as [3]:-

$$COP = \frac{T_R}{T_O - T_R} \tag{4}$$

3. Type of absorption refrigeration cycles

3.1 Single-effect machines

Most manufacturers offer single-effect machines in the range (of 100 RT to 1500 RT).

350 Kwth to 5.2 Mwth. These can be "fired" with steam at 135 to 205 kPa g (1–2 bar overpressure, 2–3 bar), corresponding to steam temperatures (110–120°C), or they can be "burned" with hot water (115–150°C). max. The pressure is 9 bar. With a coefficient of performance in the range (of 0.6–0.7), the steam consumption of a single-stage machine is about 2.3 kg/h/kwth. The required hot water flow is in the range (of 30–72 kg/h/kwth) depending on the allowable temperature drop [4].

3.2 Double-effect machines

Double-effect machines have roughly the same power range. Some manufacturers offer slightly higher minimum cooling capacities (200 RT for one company and 3500 RT for another) (700 and 1200 Kwth respectively). In the past, steam seemed to be the medium of choice for "igniting" such machines. The steam should be (8–10 bar), which corresponds to temperatures in the range of (175–185°C).

According to the information received, it is also possible to "fire" a double-effect Machine with hot water, the temperature of which should then be in the range (of 155–205°C).

The coefficient of performance, in either case, is (0.9–1.2).

The steam consumption of the double-effect machine is 1.2 kg/h per Kwth [4].

4. How absorption system works

Like the compressor in the electric vapor compression cycle, the absorption system uses its "hot" compressor (consisting of a generator, absorber, pump, and heat exchanger) to boil water vapor (refrigerant) in a lithium bromide/water solution and compresses the refrigerant vapor to a higher pressure. An increase in refrigerant pressure also increases its condensation temperature.

The refrigerant vapor condenses into a liquid at higher pressure and temperature. Since this condensation temperature is higher than the ambient temperature, heat will migrate from the condenser to the ambient air and be dissipated. The high-pressure fluid then flows through the throttle valve, reducing its pressure. Reducing its pressure also lowers its boiling point temperature. The low-pressure liquid then enters the evaporator and is cooked at a lower temperature and pressure.

Since the boiling temperature is now lower than that of the conditioned air, heat from the conditioned airflow enters the evaporator and boils this liquid. Removing heat from the air in this way causes the air to cool.

The refrigerant vapor then enters the absorber where it reverts to a liquid state It is drawn into the lithium bromide solution (absorption process).

The dilute lithium bromide solution is pumped back into the generator. Because lithium bromide (absorbent) Without boiling, the water (refrigerant) is easily separated by heating. The produced water vapor enters the condenser, the absorption solution returns to the absorber, and the process repeats.

Although the process is similar to a conventional electric vapor compression system, absorption refrigeration replaces an electric generator and an absorber called a thermal compressor, rather than an electric compressor. By using pumps instead of compressors and heat exchangers to recover heat and supply it to generators, efficiency can be increased and operating costs reduced. Double-effect absorption refrigeration adds a

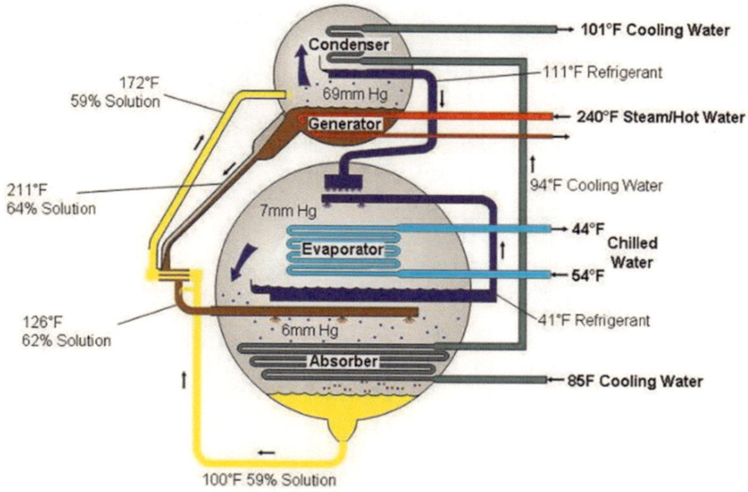

Figure 2.
Simplified diagram of a single effect absorption cycle. Source: https://gasairconditioning.com/technologies/absorption/how-it-works/

second generator and condenser to increase refrigerant flow and thus increase the cooling effect of a single-effect system for a fraction of the heat input [5] (**Figure 2**).

5. Basic thermodynamics of absorption cycle

Absorption cycles operate on the principle of refrigerant being absorbed by a chemical solution, normally water/LiBr or ammonia/water.

It is assumed that the reader is familiar with the basic absorption cycle. However, for clarity, a chiller schematic is shown in **Figure 3** the top part of the unit is the driving cycle, where heat is supplied to the generator to boil off refrigerant that is condensed in the condenser. It also concentrates the solution (say water/LiBr) and supplies it via the solution heat exchanger to the cooling sub-cycle (evaporator and absorber), so enabling it to absorb more refrigerant. The main heat exchanges with the environment are:

- High-temperature heat source: generator (i.e., steam, hot water, combustion).
- Medium-temperature heat sink: absorber and condenser (cooling tower, condenser water).
- Low-temperature heat source: evaporator (chilled water) [6].

The cycle is represented on the pressure-temperature concentration (PTX) diagram in **Figure 4** [7]. The horizontal axis indicates the solution temperature and the line inclined at 45° represents the pure refrigerant (water). From this line, the vertical saturation pressure axis values are determined. The constant concentration lines commence from the refrigerant line and indicate the percentage of LiBr. A very similar representation can be done in **Figures 3** and **5** which uses different scales: in

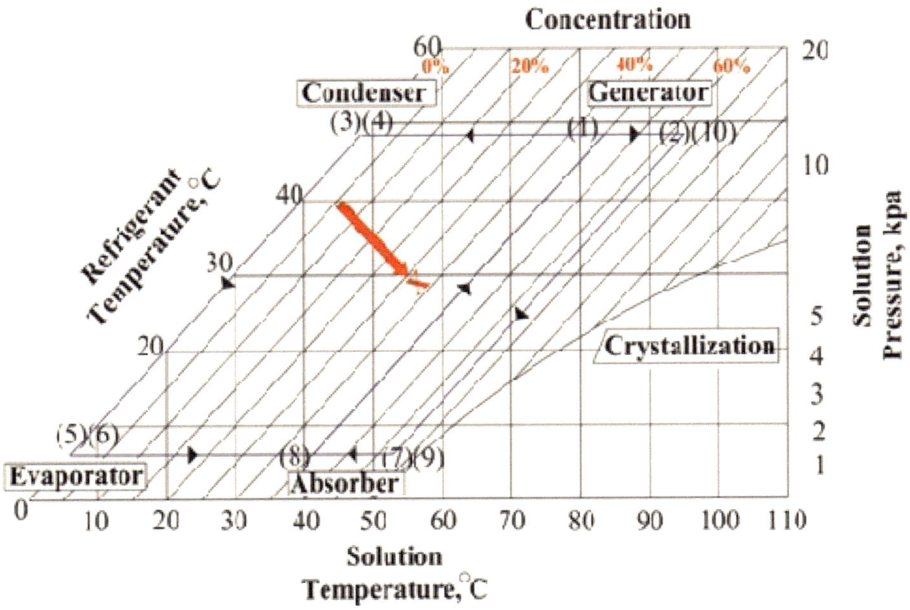

Figure 3.
Single-stage absorption cycle on PTX diagram.

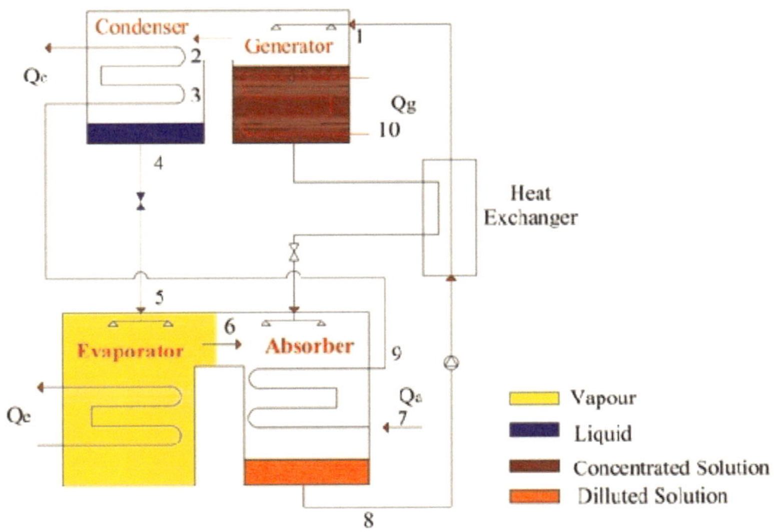

Figure 4.
Single-stage absorption chillers.

(P) and (−1/T). Two circuits (refrigerant and solution) can be identified on the PTX diagram: the left-hand side anti-clockwise quadrilateral (refrigerant circuit), and the right-hand side clockwise quadrilateral (solution circuit). It should be noted that each of the evaporation and condensation points indicates two phases: saturated liquid and vapor [8].

6. Heat-driven cooling cycles

Absorption cycles can be associated with heat-driven cooling cycles, such as the case of a heat-driven power cycle that drives a mechanical compression refrigeration system described by Herold et al., as indicated in **Figure 5**.

Heat is supplied from a high-temperature heat source (generator or boiler) and from a low- temperature heat source (evaporator), and all this heat is released to a medium-temperature heat sink (absorber and condenser). The heat flows of a generalized absorption cycle are indicated in **Figure 6**.

Although absorption chillers require small pumps to circulate/re-circulate solution and refrigerant, for ideal cycles these can be considered negligible. By applying the first law of thermodynamics to the above:

$$Q_e + Q_g = Q_a + Q_c \tag{5}$$

If the second law of thermodynamics is applied, the total generation of entropy will be zero as we are analyzing an ideal cycle.

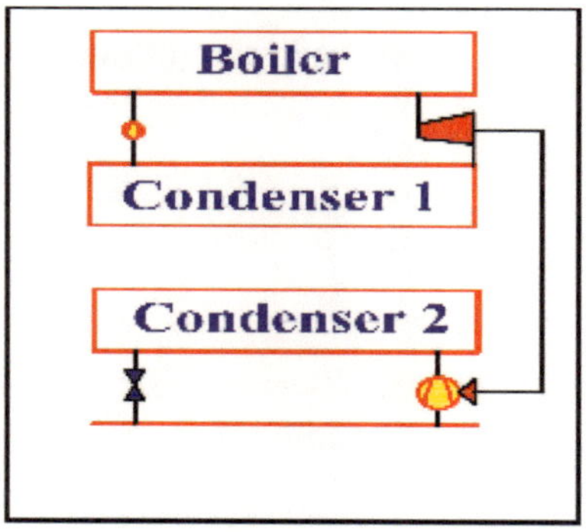

Figure 5.
Driving and cooling cycle.

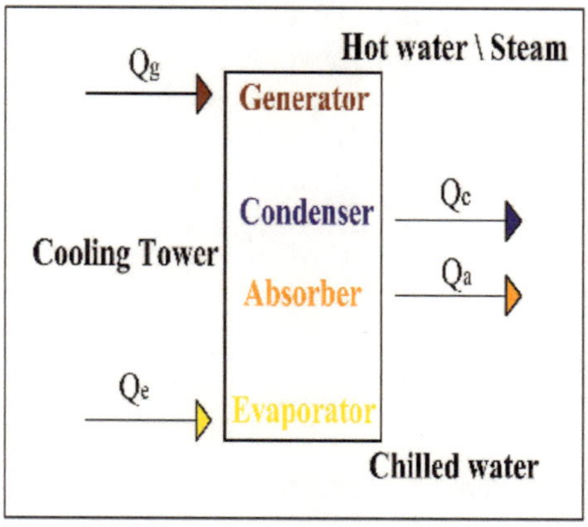

Figure 6.
Heat flows of an absorption cycle.

$$\frac{Q_e}{T_e} + \frac{Q_g}{T_g} = \frac{Q_a}{T_a} + \frac{Q_c}{T_c} \tag{6}$$

If the following assumptions are made, which are further explained later, and then the evaporator heat will be equal to the condenser heat

1. Saturated liquid-specific heat is negligible.

2. Heat of evaporation is constant.

3. Refrigerant expansion is isotropic.

4. Superheated vapor-specific heat is negligible.

$$Q_{evap} = Q_c \tag{7}$$

By combining Eqs. (5) and (6) the absorber heat will be equal to the generator heat,

$$Q_e = Q_a \tag{8}$$

If the above (7) equations are combined, the following COP equation is derived in terms of the four temperatures involved, as shown in the following Eq. (9):

$$COP = \frac{Q_e}{Q_g} - \frac{\frac{1}{T_a} - \frac{1}{T_g}}{\frac{1}{T_e} - \frac{1}{T_c}} - \frac{T_g - T_a}{T_g} \frac{T_e}{T_c - T_e} \frac{T_c}{T_a} \tag{9}$$

A way adopted by many authors has been to assume that the absorber and condenser share the same temperature. By doing this, the previous equation simplifies to the following:

$$COP = \frac{T_g - T_{ac}}{T_g} \frac{T_e}{T_{ac} - T_c} \tag{10}$$

This expression also indicates the COP as the product of the driving sub-cycle efficiency and the cooling sub-cycle COP if Carnot expressions are considered.

The two COP expressions derived above in terms of three and four temperatures are applicable to ideal heat-driven cooling cycles where the driving and cooling sub-cycles are independent of each other. This is the case for motor-driven compression systems and absorption GAX cycles. For single and multiple cycles there is no such independence between the driving and cooling sub-cycles, and further restrictions on temperatures are required and imposed [9].

7. Real absorption cycle on T-S diagram

Absorption cycles can be correctly represented on T-S diagrams, as they are best suited to represent ideal driving and cooling cycles (Carnot cycles). Due consideration should also be given to what is going to be represented on the T-S diagram. For this, it is necessary to split the refrigerant into two parts:

- The part that circulates throughout the main processes (generator, condenser, evaporator, and absorber), which is what will be represented on the T-S diagram.

- The other part, which is diluted in the solution and only circulates between the generator and absorber via the solution heat exchanger. It only experiences changes in temperature (and pressure) and is of minor thermodynamic importance if the solution heat exchanger area is sufficiently (infinitely) large. Therefore, this part of the refrigerant has no influence on the heat balance of the cycle, and will not be considered further [10].

To represent the absorption cycle on the T-S diagram, it will be made of the extended T-S diagrams developed by Eber.

A brief description of these extended diagrams will be given, but further details should be sought from the references.

A real absorption cycle (water/LiBr) is indicated in **Figure 7** where the extended T-S diagram for that mixture has been used.

Due to their different vapor pressures, no equilibrium can exist between a pure refrigerant and the solution. Therefore, the refrigerant molecules in the solution are considered, and the solution is considered indirectly. The extended curves in the vapor region (right of the saturated vapor line, indicated 58% and 64%) consider the refrigerant vapor pressure to be in equilibrium with the solution. The extended curves in the liquid region (left of the saturated liquid line, indicated 58% and 64%) consider the states of liquid refrigerant in the solution. The horizontal difference between the extended liquid and vapor lines indicates the difference of entropy required to reversibly expel 1 kg of refrigerant from a very large quantity of solution, i.e. generator heat. From the previous figures, the refrigeration circuit.

Can be identified as:

1-2-3-4-5-6-7-8.

he processes in the refrigerant cycle are:

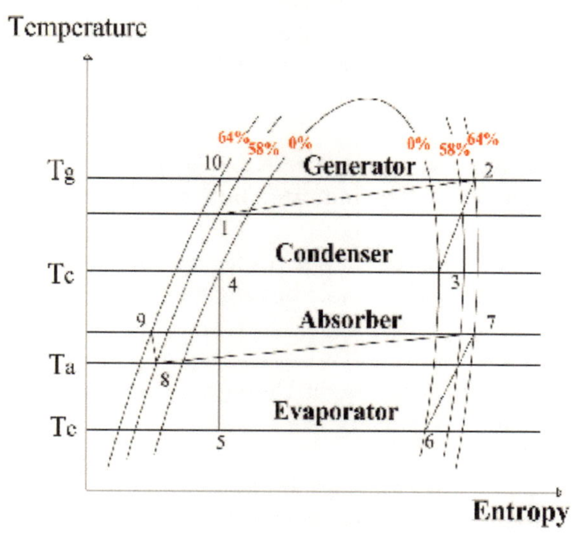

Figure 7.
Real absorption cycle on T-S diagram.

1-2: Heating and evaporation of the coolant introduced into the generator. Heat is equal to the sum of the heat of vaporization and heat of dissolution.

2-3: Cooling of the superheated steam generated in the constant pressure generator down to the saturation temperature Tc.

3-4: Condensation.

4-5: Isenthalpic expansion of the refrigerant from the condensation to the evaporation temperature.

5-6: clutch.

6-7: Isobaric (constant pressure) heating of the refrigerant in the form of saturated vapor or superheated vapor.

7-8: Absorption of superheated steam (state 7), in which the heat of solution and the heat of vaporization are released.

A minimum number of 12 idealizing assumptions are required to obtain an ideal absorption cycle. A summary of the assumptions follows:

1. Saturated liquid specific heat is negligible: The specific heat of the saturated liquid refrigerant along the lower boundary line is equal to zero. The implication of this is that the saturated liquid line is vertical on the T-S diagram.

2. Heat of evaporation is constant: The heat of evaporation is constant and independent of the temperature. This defines the saturated vapor line as a hyperbola on the T-S diagram.

3. Refrigerant expansion is isotropic: If this process changes from isenthalpic to isotropic, it becomes vertical on the T-S diagram (process 4-5).

4. Superheated vapor specific heat is negligible: The specific heat at a constant pressure of the superheated vapor is equal to zero. Therefore, the isobaric heating and cooling of vapor are represented as vertical lines on the T-S diagram.

5. Heat of solution varies only with concentration: The heat of solution/only depends on the composition of the solution and is not influenced by the temperature. Therefore, a family of hyperbolic curves will be derived for each solution concentration.

6. Solution-specific heat is negligible: Therefore, both the cooling and heating processes of the solution will be indicated by vertical lines on the T-S diagram.

7. Heat exchanger area is infinitely large: If the solution heat exchanger is sufficiently large, there is no sensible heating and cooling in the generator and absorber.

8. Solution circulation is infinitely high: Therefore, the concentrations of the strong and weak solutions converge.

9. Entropy of solution mixing is negligible with respect to the evaporation heat: This means that the mixing of solution will tend to be reversible.

10. Refrigerant vapor is an ideal gas: Which means that it complies with the properties of ideal gases.

11. There is no absorbent vapor pressure: Therefore, only the refrigerant will depart from the solution in the generator, and rectification is not necessary.

12. The solution pump work is considered to be negligible [11].

8. Working pairs in absorption refrigeration units

The working fluid pairs to be used in absorption refrigeration units must be selected on the bases of their suitability under the conditions of Their thermodynamic properties, chemical and thermal stability, and toxicity. Examples of working substance pairings and applications that have been extensively developed are:

A. Ammonia water is used in air conditioning and refrigeration. Ammonia is a refrigerant and water is a solvent. The performance values of the process depend on the temperature, which reaches a level of (−45°C). The main field of application of ammonia is large-scale use in the MW range.

B. The ammonia-hydrogen combination has been used in domestic refrigerators.

C. Aqua-LiBr solution has been used in domestic and commercial air conditioners, which can reach power values of 0.65 (single stage) and 1.10 (two stages), the required heating temperature is 90–115° C, and the small volume reaches the temperature level (6° C).

Usually, one of two types of absorption chillers are used, mainly depending on the required cooling temperature:

A. For the cooling temperature above 5°C, a water/lithium bromide absorber is the most commonly used and must be water-cooled.

B. The ammonia water machine can be used when the cooling temperature is lower than 5°C. It can be air-cooled or water-cooled.

A pair of thermodynamic properties used in absorption refrigeration plants are:

- The latent heat of vaporization of the refrigerant is high,

- Refrigerant vapor pressure is low but above atmospheric pressure,

- Thermal mixing or low or negative reaction,

- The intrinsic heat capacity of the absorbing solution is low compared to the latent heat of the refrigerant,

- no crystallization or freezing within the range of operating conditions, and.

- High solubility of the refrigerant in the absorber [11].

9. Conclusion

1. If the vapor leaving the evaporator is accompanied by some liquid, this is not a major disadvantage in the operation of absorption chillers, but it can be a risk in compression chillers. To prevent this (the vapor escaping with the liquid) in compression refrigeration systems, the compressor must have special controls. This issue can be very important in applications with widely varying cooling requirements.

2. Large-capacity absorption chillers take up more space than compression chillers. However, the units can also be placed outside the building and can be made taller rather than wider and deeper, taking up less floor space.

3. With the development of urban natural gas absorption chillers, the consumption of natural gas has increased year by year, resulting in air pollution, especially in summer.

4. Some industries generate unwanted heat through various processes, which is wasted by being transferred to the surrounding air. However, these industries actually need to cool down. Using waste heat in an absorption chiller to generate the required cooling is a viable solution and, in this case, establishes a viable symbiosis.

5. Absorption chillers have a very low COP, using about four times as much energy (in the form of heat) to produce a given amount of cooling.

6. The cooling of the condenser and absorber in absorption chillers uses about twice as much water as the condenser in compression chillers (if these systems are water-cooled). Absorption chillers are not widely used due to the scarcity of water and the need to conserve water as its value increases every day.

Author details

Ali F. Ali Fadiel* and Yoonis A.M. Esham
Department of Mechanical Engineering High Institute of Technical Sciences, Tobruk, Libya

*Address all correspondence to: dr_ali.f@hotmail.com

References

[1] Nikbakhti R, Wang X, Hussein AK, Iranmanesh A. Absorption cooling systems–Review of various techniques for energy performance enhancement. Alexandria Engineering Journal. 2020; **59**(2):707-738

[2] Wang S. Handbook of Air Conditioning and Refrigeration. New York: McGraw-Hill; 1984

[3] ASHRAE Handbook of Fundamentals. Atlanta: American Society of Heating, Refrigeration and Air-Conditioning Engineers; 2005

[4] Felli M. Absorption Refrigeration Thermodynamics. Parl 1A ASHRAE Transactions. 1999;**89**(2748):189-204

[5] Alefeld G. Double Effect, Triple Effect and Quadruple Effect Absorption Machines. 16th Int. Vol. 2 Paris: Congress of Refrigeration; 1999. pp. 951-956

[6] ASHRAE Handbook of HVAC Applications. Atlanta: American Society of Heating, Refrigeration and Air-Conditioning Engineers; 2004

[7] Lahoud C, Al Asmar J, Brouche M. Review of cogeneration and trigeneration systems. African Journal of Engineering Research. 2018;**6**(3): 39-54

[8] An Introduction to Air Conditioning Systems. Harwell UK: Good Practice Guide No 256, Energy Efficiency Best Practice Program, ETSU; 1999

[9] Eber N. A New Analysis of Rectification in Absorption Refrigeration. Madrid: paper 3.14, Proceedings of the XII International Congress of Refrigeration; 1967. pp. 1339-1351

[10] Herrmann F. The absorption refrigerator as a thermal transformer. European Journal of Physics. 2009; **30**(2):331

[11] Tozer, James R. Thermodynamics of Absorption Refrigeration (Ideal Cycles), Paper E5, International Absorption Heat Pump Conference. Louisiana, USA: ASME pub.; 1994. pp. 393-400

Chapter 2

Fin and Tube HVAC Evaporators Design by the Hybrid Method

Giuseppe Starace, Silvia Macchitella and Gianpiero Colangelo

Abstract

The hybrid method allows the overall performance of a fin and tube evaporator to be determined, beginning from local analysis results and combining the accuracy of the data obtained with a small-scale numerical approach with the low processing costs of the analytical approaches. The program calculates heat transfer rates, pressure drops, and temperature fields for both sides of the heat exchanger using regression equations derived from known data (analytical, experimental, or numerical). The hybrid method has been progressively refined and enhanced with the goal of modeling heat exchangers more closely to actual typologies often used for HVAC applications that involve intricate circuit arrangements. In order to aid designers to choose the optimal configuration, various refrigerant circuitry layout choices were examined as well as a proper trade-off analysis was performed.

Keywords: heat exchanger design, hybrid method, evaporator, circuit arrangement, refrigerant circuit layout, refrigeration

1. Introduction

In order to save production and operating costs, heat exchangers (HXs) employed for refrigeration application require excellent design and optimization procedures. The design process should be efficient and accurate at the same time, avoiding oversizing that is frequently the result of poor design accuracy and that consequently causes production costs to rise. On the other hand, the higher precision obtained with CFD approaches can significantly slow down the optimization process while still resulting in high design costs.

Corberán et al. [1, 2] developed a distributed model for finned exchangers acting as evaporators and condensers using a control volume approach. This model iteratively calculates the pressure and temperature characteristics on the air and refrigerant side while also incorporating mass, energy, and momentum equations.

Jiang et al. [3] created *CoilDesigner*, a general-purpose modeling and design tool for air-to-refrigerant heat exchangers that use a "segment-by-segment" methodology. A basic finned-tube coil with complex refrigerant circuitry layout, many working fluids, and no restrictions on the number and positioning of the intake and outlet streams was used for the research.

Tarrad and Al-Nadawi [4] developed a mathematical model utilizing the segment-by-segment method to forecast the performance of a louvered finned-tube evaporator operating with pure and zeotropic refrigerants. It was demonstrated that there was a good correlation between the numerical results and the data coming from the experimental testing done on an air conditioner.

Due to the large number of design factors, the optimization process for a finned-tube HX can be extremely difficult. It has been also demonstrated that, due to constraints that frequently arise in small installation spaces or other manufacturing issues, optimizing the refrigerant path by altering circuit arrangement is the best method for cost savings, as discussed by Yun and Lee [5] and Matos et al. [6]. Some researchers investigated the effect of refrigerant circuit layout by conducting tests with various circuitry layouts, such as Joppolo et al. [7], who performed a numerical study on a fin and tube condenser, estimating the heat transfer rate between air and refrigerant using the ε-NTU method for each element into which the condenser geometry was divided.

Sim et al. [8] and Wang et al. [9] have recently carried out experiments on finned-tube heat exchangers with reversely variable circuitry to improve performance. When a heat pump is used for both heating and cooling, the traditional FTHXs have two-way fixed circuitry with the same refrigerant flow channel in the opposite direction. In tests conducted by Wang et al. [9], it was found that allowing the exchanger's circuitry arrangement some flexibility - depending on whether it was acting as an evaporator or condenser - increased the heat pump's overall performance.

Starace et al. [10] created an alternative design procedure called the *hybrid method* using a multi-scale approach, starting from data sets coming from either analytical or numerical studies or also from experimental investigations in order to reduce computational efforts while achieving a high accuracy of the results. The hybrid approach was originally used on compact cross-flow HXs, where the entire geometry was split into a number of control volumes, each of which had a cold side and a warm side. Through the application of a regression technique, Carluccio et al.'s thermofluidynamic simulation findings on the two finned surfaces of the HXs [11] were used to develop the heat transfer prediction functions, extending the local results throughout the full geometry of the HX.

By using small-scale experiments as a starting point, Fiorentino and Starace [12] created another hybrid technique application for countercurrent evaporative condensers in order to assess their performance. Results indicate that, if compared to experimental tests, the method is accurate and can determine the air temperature and relative humidity at the output with errors of 2.5% and 4%, respectively. Then, Starace et al. [13] applied the method to a plate-finned evaporator with a simple refrigerant circuit configuration that was organized into independent rows and fed uniformly at the top of the exchanger. Moreover, Starace et al. [14] achieved progress in the development of the hybrid technique by applying it on evaporators with complex circuit layouts to assess the impact of circuitry configuration on the overall performance in terms of heat transfer rate and refrigerant pressure decreases. Afterward, the hybrid approach algorithm underwent additional modifications to make it even more adaptable and compatible with real heat exchangers [15]. Part of the algorithm code was altered in order to broaden the model's application range, and new tests were conducted by changing the operating conditions.

Here, various simulations using the hybrid method on an evaporator with different circuitry layouts together with a trade-off analysis have been carried out to assess the influence of refrigerant circuit configurations on performance while taking into account the effects of two-phase fluid.

2. The hybrid method

The staggered finned-tube exchanger used here works with a refrigerant that evaporates due to heat transfer, while air flows through the fins normally to the tubes. Two or more complex circuits with the same number of pipes each - within which the refrigerant flows - constitute the evaporator. Since the curves connecting the pipes are ignored throughout the heat transfer process, they are regarded as adiabatic (future developments are planned to provide an estimation of the heat losses to be validated with an experimental campaign).

Each of the tube-centered elementary cells into which the entire geometry of the HX is divided is identified by the model using a three-dimensional matrix (**Figure 1**).

An edge cell without any pipes is positioned at the bottom of each row of odd numbers and at the top of each row of even numbers in the staggered pipes of the evaporator. The properties in the border cells are calculated independently. The heat

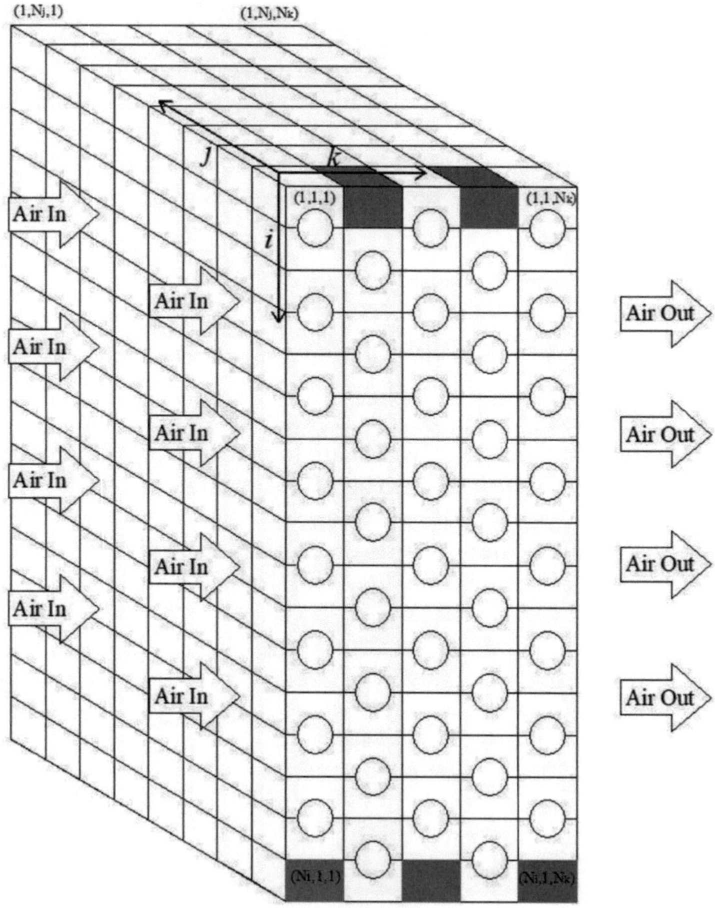

Figure 1.
Schematic representation of the evaporator by a three dimensional matrix [16].

transfer rate and wall temperature for all other cells are computed using an iterative process that requires the following input data:

- the HX geometry;
- the arrangement of the refrigerant circuits;
- the operating conditions;
- the regression coefficients, which are determined by applying the outcomes of experimental, numerical, or analytical research.

Following entering the data, the algorithm determines the refrigerant flow direction, the circuit to which the tube belongs, as well as the branch tube (**Figure 2a**) or the previous tube in the refrigerant flow order, in order to assign to each cell a refrigerant flow rate and a vapor quality both consistent with the circuit under consideration. The flow rate and vapor quality are set equal to the input values in the first delivery cell of each circuit. For all the subsequent cells, the algorithm assigns the same refrigerant characteristics at the branch pipe's outlet, assuming that the curve pipe sections are adiabatic. The initial row of the HX is thought to have an evenly distributed airflow; however, the distribution in each of the succeeding rows is created by mixing the air from the cells of the prior rows (**Figure 2b**).

As long as the convergence condition between the heat transfer rate on the air side and that on the refrigerant side is verified, taking into account the convective contribution of the air and refrigerant and the conductive contribution of the piping, the algorithm iteratively calculates the wall temperature of the pipe for each cell of the HX, as reported in Eq. (1).

$$\dot{Q}_R = \dot{Q}_A \tag{1}$$

Correlations in **Table 1** provide the additional thermodynamic variables as a function of the inner and outer wall temperatures.

The heat transfer rates of the edge cells are calculated as a percentage of the neighboring cells' heat transfer rates, which are found below the edge cell in odd ranks and above in even ranks (**Figure 2**), instead of being obtained by the convergence method.

Finally, the algorithm verifies pressure drops on the air side and refrigerant side once all cells have been calculated as in Eq. (2).

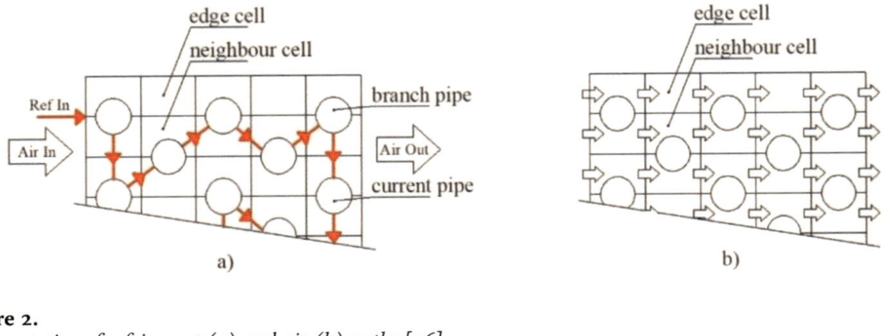

Figure 2.
Representation of refrigerant (a) and air (b) paths [16].

Air side	Reference	Refrigerant side	Reference
Heat transfer coefficient	[17]	Total heat transfer	[23, 24]
Lewis number	[18]	Pressure drop	[25]
Overall fin efficiency	[19]		
Pressure drop	[20]		
Air specific humidity	[21]		
Friction factor	[22]		

Table 1.
Correlations for calculation of air side and refrigerant side variables.

$$\Delta p_A(i,j) = \sum_{k=1}^{N_k} \Delta p_A(i,j,k) \quad (2)$$

If the condition in Eq. (3) is not satisfied, the algorithm then redistributes the air mass flow rate into each cell again, changing the flow rate according to the error from the mean value.

$$\Delta p_A(i,j) = \Delta p_{A,m} \quad (3)$$

$$\Delta p_{R,z} = \Delta p_{R,m} \quad (4)$$

As in Eq. (4), the pressure drops for each circuit of the refrigerant should be within 1% of the mean value. The program then distributes the refrigerant flow rates among the circuits until each *zth* circuit's condition is confirmed. The conservation of the total input mass flow rate is guaranteed throughout the whole calculation, when changing both the local air and refrigerant flow rates.

The predictive function for calculating the heat power and all the thermodynamic parameters involved is extrapolated through a quadratic regression method with the goal of finding the correlation between the input and output variables on both the air and refrigerant sides. Here, the data used to perform the regression technique were derived from the experimental correlations shown in **Table 1**. Due to the model's great flexibility, it is always possible to use available numerical analysis in place of them. While the air side response variables $T_{A,outt}$, $i_{A,out}$, and $p_{A,out}$ are determined using Eq. (5), the refrigerant side response variable ${}_t x_{R,out}$ is calculated as in Eq. (6).

$$\beta \cdot \alpha^T \quad (5)$$

$$\beta \cdot \gamma^T \quad (6)$$

where the 15 polynomial coefficients derived from the regression analysis are the elements of the vector β, and α and γ are the vectors whose elements are displayed in **Tables 2** and **3**, respectively.

3. Simulation setup

Due to the large number of design factors, the optimization process is quite difficult when facing the design of a HVAC evaporator. For instance, spreading the

Element	Value	Element	Value	Element	Value
α_0	1	α_5	$G_R x_{R,in}$	α_{11}	$T_R T_{w,i}$
α_1	G_R	α_6	$G_R T_{w,i}$	α_{12}	G_R^2
α_2	$x_{R,in}$	α_7	$G_R T_R$	α_{13}	$x_{R,in}^2$
α_3	$T_{w,i}$	α_8	$x_{R,in} T_{w,i}$	α_{14}	$T_{w,i}^2$
α_4	T_R	α_9	$x_{R,in} T_R$	α_{15}	T_R^2

Table 2.
Components of vector α [16].

Element	Value	Element	Value	Element	Value
γ_0	1	γ_5	$T_{A,in} RH_{A,in}$	γ_{11}	$V_{A,in} T_{w,o}$
γ_1	$T_{A,in}$	γ_6	$T_{A,in} V_{A,in}$	γ_{12}	$T_{A,in}^2$
γ_2	$RH_{A,in}$	γ_7	$T_{A,in} T_{w,o}$	γ_{13}	$RH_{A,in}^2$
γ_3	$V_{A,in}$	γ_8	$RH_{A,in} V_{A,in}$	γ_{14}	$V_{A,in}^2$
γ_4	$T_{w,o}$	γ_9	$RH_{A,in} T_{w,o}$	γ_{15}	$T_{w,o}^2$

Table 3.
Components of vector γ [16].

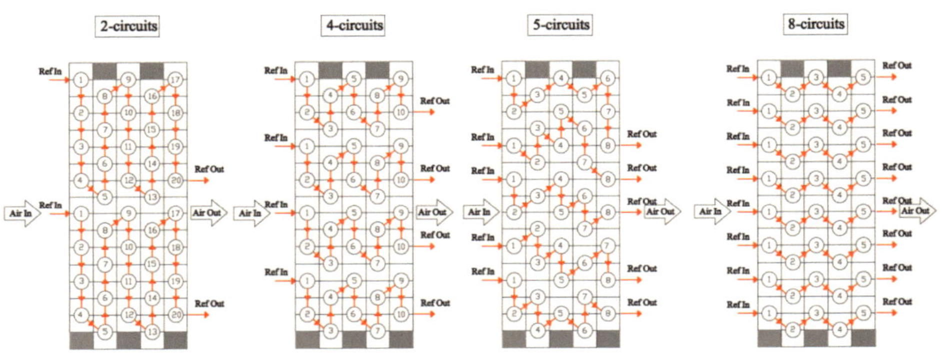

Figure 3.
Circuitry arrangements of Set 1 *[16].*

refrigerant flow across two or more circuits will minimize refrigerant pressure drops when large flow rates are required.

In order to explore the impact of circuit layout on the HX performance in terms of heat transfer rate and refrigerant pressure drops, four distinct simulation tests using three different circuitry designs have been carried out in the present work. There are four different layouts for Set 1, which is represented in **Figure 3**. All of the circuit inlets are located on the same side of the heat exchanger, while the outlets are situated on the opposite one. The air intake - that flows over the fins normally to the axes of the pipes - is positioned on the same side as the refrigerant inlets. In Set 2 (**Figure 4**), instead, the air inlet is located on the same side as the refrigerant outlets. The three circuitry arrangements from Set 3 (**Figure 5**) consist of four-circuit layouts with more pipes per row (12 tubes per row in Set 3, 8 tubes per row in Set 1 and Set 2).

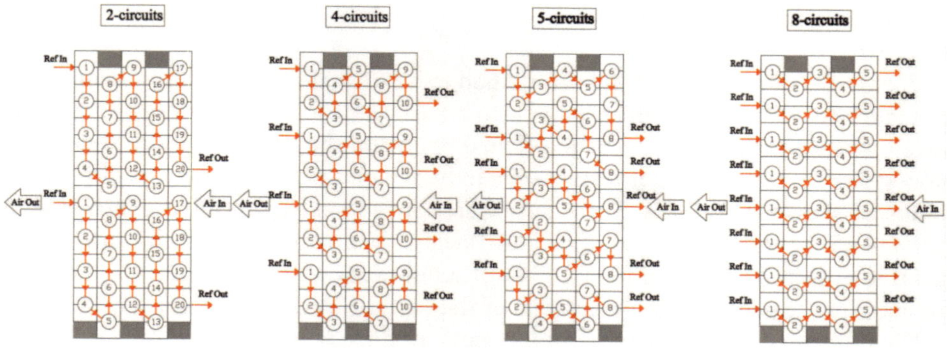

Figure 4.
Circuitry arrangements of Set 2 *[16].*

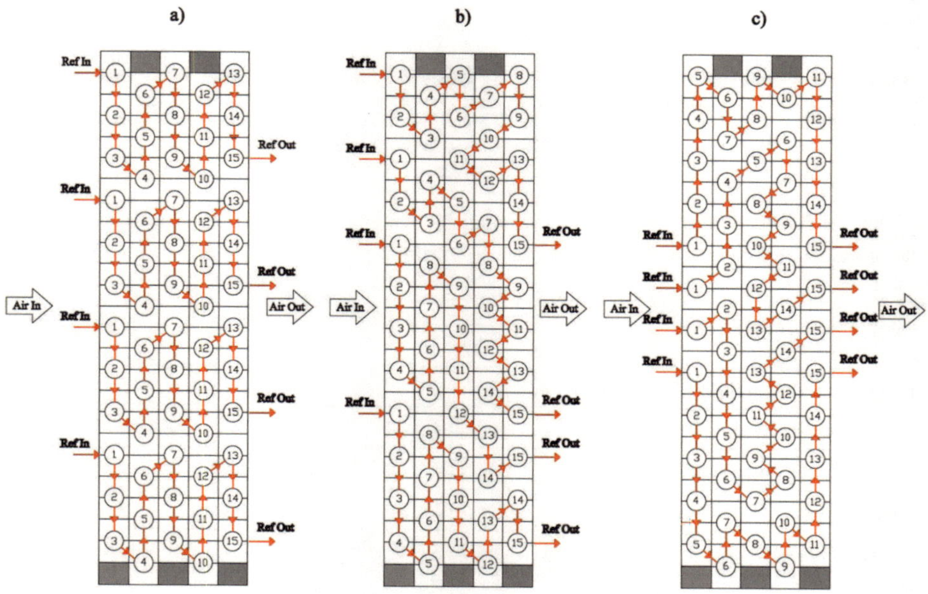

Figure 5.
Circuitry arrangements of Set 3 *[16].*

Each test had a clear objective:

- investigation on the circuitry arrangements of Set 1 with various refrigerants (test a);
- investigation on the circuitry arrangements of Set 1 with various refrigerant flow rates but the same heat transfer rate (test b);
- comparison between the circuitry layouts of Set 2 and Set 1 (test c);
- investigation on the circuitry layouts of Set 3 made up entirely of 4-circuit layouts to help designers in optimizing evaporator circuit arrangement (test d).

4. Results and discussion

The results of running the hybrid method to a 5-row evaporator with various circuit arrangements are reviewed in terms of heat transfer rate and refrigerant pressure drops as well as compared through a trade-off analysis. With the aim of assisting designers in making design decisions, four tests were conducted on three different sets of configurations, as shown in **Table 4**. These tests were carried out by varying the refrigerant (test a) and the refrigerant mass flow rate, keeping the same heat transfer rate (test b), the air inlet side (test c), and the number of tubes per row (test d). **Table 5** provides a summary of the geometrical parameters of the tested HVAC evaporator.

The purpose of test a was to compare performance while varying the number of circuits. **Figures 6** and **7** illustrate variations in heat transfer rate and refrigerant pressure drop as a function of the number of circuits, respectively, for cases a) with R134a, R410a, and R32 and b) with R404a, R507a, R1234yf, and R1234ze. Circuitry layouts from Set 1 were used for test a, as illustrated in **Figure 3**.

Quantity	Unit	Test *a*	Test *b*	Test *c*	Test *d*
Refrigerant fluid	—	a) R134a/R410a/R32 b) R404a/R507a/ R1234yf/ R1234ze	R134a	R32	R32
Number of tubes per row	—	8	8	8	12
Refrigerant mass flow rate	kg/s	a) 0.047 b) 0.058	0.047/0.103/ 0.122/0.195	0.036	0.047
Air mass flow rate	kg/s	a) 0.644 b) 0.515	0.644	0.644	0.940
Evaporation temperature	K	271.5	271.5	271.5	271.5
Air inlet temperature	K	288	288	288	288
Air inlet relative humidity	—	0.65	0.65	0.65	0.65
Inlet vapor quality	—	0.2	0.2	0.2	0.2
Air inlet velocity	m/s	a) 5 b) 4	5	5	5

Table 4.
Simulation test setup for test a, *test* b, *test* c, *and test* d *[16].*

Tubes			Fins		
Quantity	**Unit**	**Value**	**Quantity**	**Unit**	**Value**
Material	—	Copper	Material	—	Aluminum
Internal diameter	mm	7.38	Thickness	mm	0.1
External diameter	mm	7.94	Pitch	mm	2
Length	mm	500			
Longitudinal pitch	mm	21.65			
Transversal pitch	mm	25			

Table 5.
Geometrical parameters of tested HVAC evaporator [16].

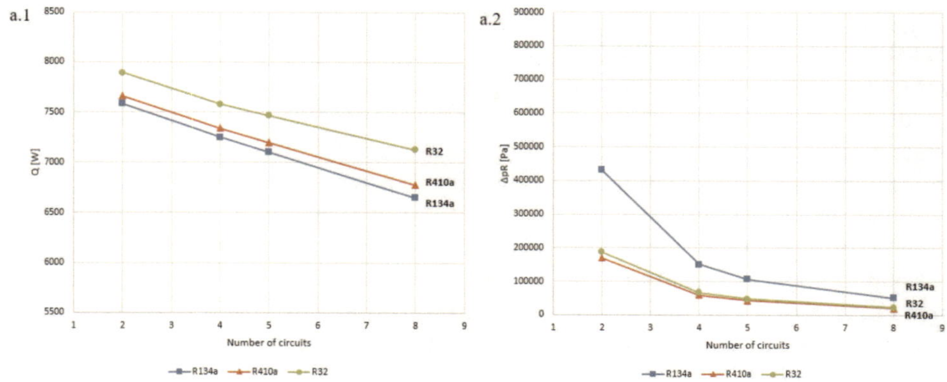

Figure 6.
Test a, case a): Heat transfer rate (a.1) and refrigerant pressure drops (a.2) for each of the four circuit layouts of Set 1. Tested refrigerants: R134a, R410a, and R32 [16].

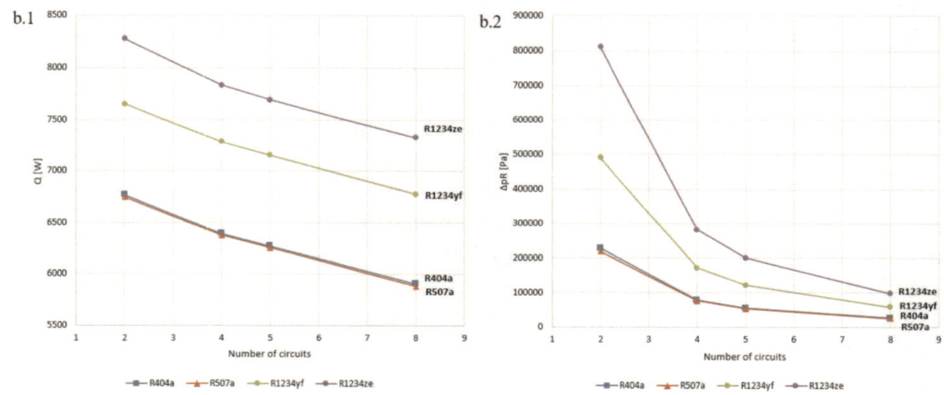

Figure 7.
Test a, case b): Heat transfer rate (b.1) and refrigerant pressure drops (b.2) for each of the four circuit layouts of Set 1. Tested refrigerant: R404a, R507a, R1234yf, and R1234ze [16].

Results showed that when the number of circuits under consideration rises, the heat transfer rate drops approximately linearly. For all the refrigerants taken into consideration, there is a mean reduction in heat transfer rate for the 8-circuit configuration of 11.14% for case a) and 12.12% for case b) compared to the 2-circuit arrangement. A drop in the refrigerant flow rate through each circuit caused by an increase in the number of circuits also brings a reduction in the convective heat transfer coefficient, which has an adverse impact on the overall heat transfer. As a result, the 8-circuit layout's outlet vapor quality was shown to be worse than the 2-circuit configuration. On the other hand, as the number of circuits increases, flow rate across each circuit decreases, resulting in a strong decreasing of pressure drops with a parabolic pattern for all the refrigerants investigated.

R32 appears to be the most efficient refrigerant in case a) because it has the highest heat transfer rate and almost the smallest pressure drops. The same cannot be said for R1234ze, which has the highest heat transfer rate among case b) refrigerants but also exhibits the highest pressure drops. R134a, on the other hand, performs the worst, with the slowest rate of heat transfer and the greatest pressure reductions.

The R134a refrigerant was used in test *b* on Set 1 circuitry design (**Figure 3**), with the goal of comparing various layouts at the same heat transfer rate in terms of refrigerant pressure drops and outlet vapor quality by adjusting the input refrigerant mass flow rate, as stated in **Table 4**. Results in **Figure 8** demonstrate that the 8-circuit configuration, compared to the 2-circuit configuration, has a smaller refrigerant pressure drop at the same heat transfer rate. Therefore, a coil designer could decide for an 8-circuit arrangement to lower operating costs while retaining the same heat transfer rate performance by increasing the refrigerant flow rate. In test *b*, it is demonstrated that by using an 8-circuit layout rather than a 2-circuit one, it is possible to reduce pressure drops by 45.3% while maintaining the same performance level.

Another effect of selecting a layout with many circuits is that, in order to achieve a given heat transfer rate, the total flow rate must be increased because of the lower local velocities and lower mean outlet vapor quality (in **Figure 8b**, the vapor quality decreases from 0.99 in the 2-circuit configuration to a value of 0.35 in the 8-circuit configuration (−64%)). Here, the flooded evaporator principle is evident, which denotes the recirculation of the liquid phase and ensures that a saturated vapor exits the evaporator toward the compressor.

test *c* was performed to compare the circuitry designs of Sets 1 and 2. The only difference between the two sets is that Set 1's air inlet is situated on the same side as the refrigerant inlets, while Set 2's air inlet is located on the side with the refrigerant outlets. Both sets have the same refrigerant path and the same number of circuits. R32 served as the working fluid in test *c*, and the inlet conditions are listed in **Table 4**. Results in **Figure 9** demonstrate how, for Set 2 circuitry layouts, the heat transfer rate linearly reduces as the number of circuits grows (a). However, it is evident from comparing Sets 1 and 2 that the air inlet side has little to no impact on the HX's heat transfer rate (**Figure 9a**). In any scenario, Set 2 performs worse than Set 1 with a 0.74% lower average deviation and 6.78% fewer refrigerant pressure drops than Set 1. Finally, the refrigerant pressure drop follows the same parabolic pattern as tests *a* and *b*: as the number of circuits rises, the pressure drop drastically declines (**Figure 9b**).

In order to investigate the effects of the refrigerant path on the coil performance while maintaining a consistent number of circuits, test *d* was performed using R32 on the three configurations of Set 3. This is the main reason why Set 3's HX has more tubes per row than Set 1's and Set 2's.

For the investigated scenarios a), b), and c) of Set 3 (see **Figure 10**), there was no discernible effect of the refrigerant path on heat transfer rate. Actually, the largest divergence of the heat transfer rate from configuration a) to b) is −0.03%.

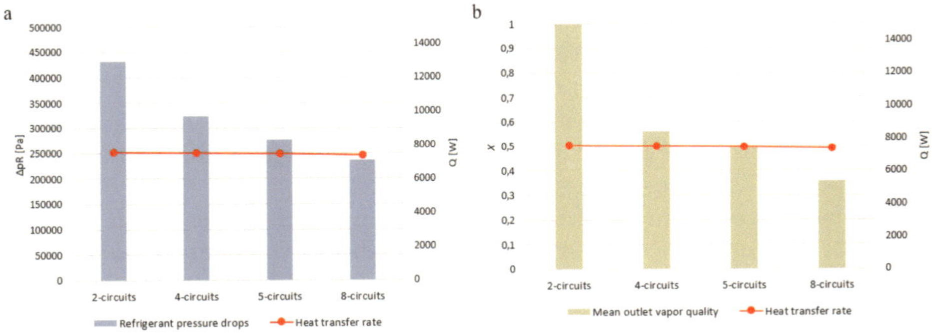

Figure 8.
Test b*: Refrigerant pressure drops (a) and mean outlet vapor quality (b) for each of the four circuit layouts of Set 1 [16].*

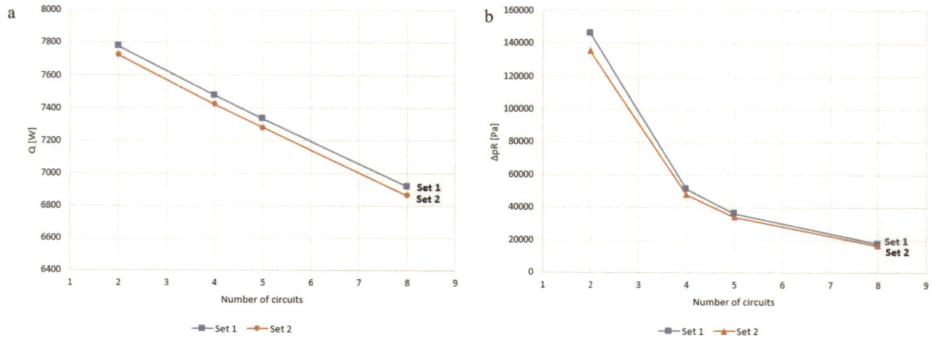

Figure 9.
Test c: *Heat transfer rate (a) and refrigerant pressure drops (b) for each of the four circuit layouts of* Set 1 *and* Set 2 *[16].*

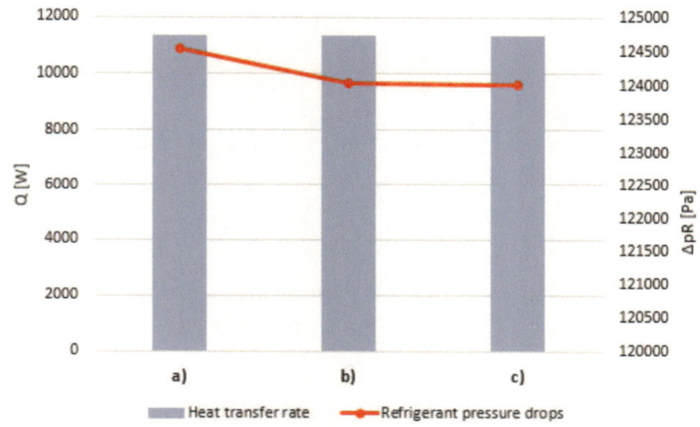

Figure 10.
Test d: *Heat transfer rate and refrigerant pressure drops for circuitry arrangement a), b), c) of* Set 3 *[16].*

5. Trade-off analysis

Pressure loss throughout a refrigeration unit's whole circuit as well as the change in LMTD brought on by the drop in saturation pressure in both the evaporator and the condenser have an impact on compressor power. As a result, while maintaining the set constraints, the heat exchanger selection is a crucial component of the design of the entire refrigeration unit [26].

Diagrams in **Figures 11** and **12** illustrate the trade-off between *UA* and Δp_R for multiple refrigerants and per number of circuits, extrapolating data from the test findings presented in Section 3. The same HX geometry and either air or refrigerant inlet conditions were adopted for all testing; however, alternative refrigerant paths were used. Each diagram's point represents a single tested circuitry arrangement (set 1, **Figure 3**). It must be emphasized that even if an optimization of the refrigerant path with the same number of circuits is feasible, as proved with test *d* and as stated in Section 3, the expected modifications in terms of *UA* and pressure drop would be quite minimal. As a result, it is safe to conclude that the diagrams in **Figures 11** and **12** are indicative of the actual evaporator performance with respect to a specific number of circuits.

The 8-circuit arrangement results in a modest reduction in *UA* as well as in a significant decrease of its pressure drop when compared to the 2-circuit layout. All

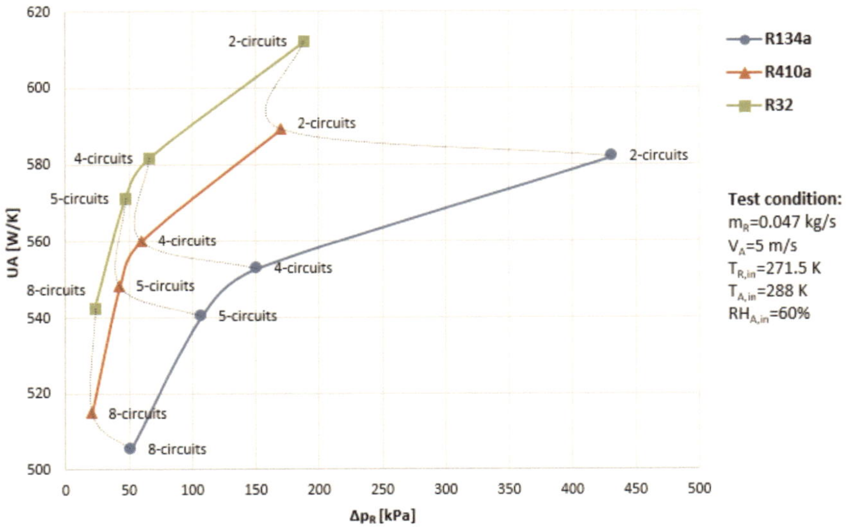

Figure 11.
UA *vs. refrigerant pressure drop diagram for refrigerants R134a, R32, and R410a [16].*

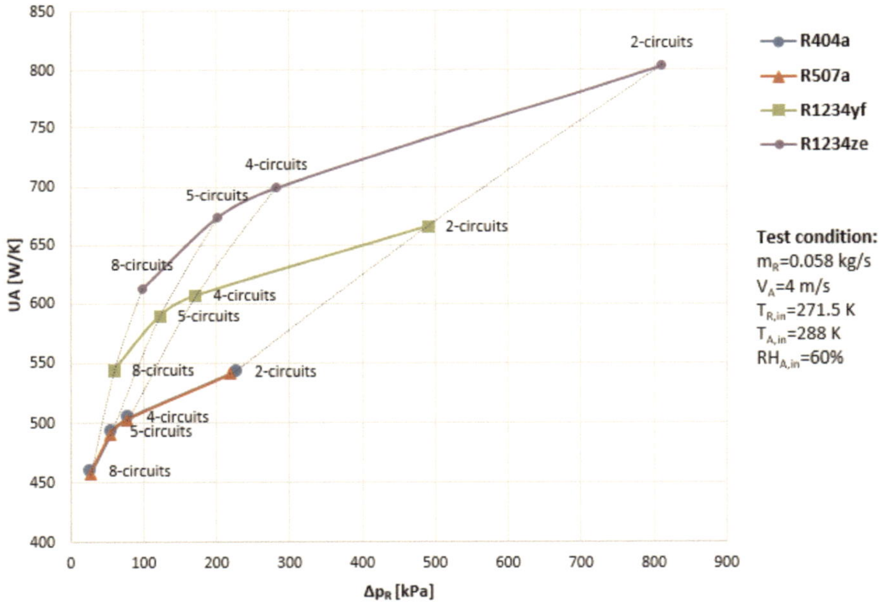

Figure 12.
UA *vs. refrigerant pressure drop diagram for refrigerants R404a, R1234yf, R1234ze, and R507a [16].*

refrigerants exhibit the same trend but with varied curve slope due to the distinctive characteristics of each refrigerant.

When all refrigerants are taken into account and the 8-circuit arrangement is compared to the 2-circuit one, the pressure drop variation range is between 87.63% and 88.05%, with the minimum and maximum changes in *UA* of 11.36% and 23.69%, respectively. R32 exhibits the best behavior of the evaluated refrigerants from this perspective. With R32, the pressure drop can be reduced by 87.63% when switching from a 2-circuit layout to an 8-circuit one, even with a slight loss in UA of 11.36%.

However, using R1234ze results in a more severe decrease of UA, equal to 23.69%, in exchange for a reduction in pressure drop of 87.89% when an 8-circuit design is used on an evaporator rather than a 2-circuit configuration.

6. Conclusion and future developments

A new feature has been added to the previous version of the hybrid method algorithm, aimed to address specific characteristics of plate-finned tube HVAC evaporators with complex circuit layouts. In order to provide designers with meaningful information for the design process, the multi-scale model, which conducts a local analysis to acquire the heat transfer parameters on each elementary volume, was implemented to compare different circuitry layouts. The regression technique was applied to data coming from experimental correlations found in the literature and useful to set up the prediction functions, then used to calculate the thermodynamic parameters both at the refrigerant and air sides. Three sets of circuits were subjected to four distinct simulation tests. The results showed that for all of the investigated refrigerants, heat transfer rate reduces approximately linearly as the number of circuits is increased, but refrigerant pressure drops substantially decrease with a parabolic pattern (R134a, R410a, R404a, R32, R507a, R1234ze, and R1234yf). Other tests with R134a on various circuitry configurations at the same heat transfer rate revealed that an 8-circuit configuration can be chosen to increase the refrigerant flow rate by 314% while maintaining the same performance in terms of heat transfer rate. This will reduce the refrigerant pressure drop by 45.3% and, consequently, operating costs. Additionally, vapor quality degrades when the liquid phase can be recycled through the evaporator to ensure that the fluid arrives at the compressor as saturated vapor, as is flooded evaporators.

Additional research on alternative circuit layouts revealed that the air inlet side had little to no impact on the HX's performance in terms of heat transfer rate. Nevertheless, designs with the air inlet located on the side opposite the refrigerant entrances performed worse than those on the same side, with an average heat transfer rate deviation of 0.74% less and a reduction in refrigerant pressure drops of 6.78%. Finally, other experiments also revealed that optimizing the refrigerant path is technically feasible while maintaining the same number of circuits, although the advantage in terms of improved performance is low.

In order to provide designers with a standard procedure to evaluate HVAC evaporators' performance when the circuitry layout varies, trade-off curves illustrating *UA* versus refrigerant pressure drop were created, using the data extracted from all the tests.

Further activities are planned in order to improve the algorithm of the hybrid method by adding various features such as the modeling of the frost formation and growth on the coil's surface and a more closer-to-reality mixing of the air in the cells of the ranks following the first one. Furthermore, experimental investigations are scheduled to test the heat losses through the pipe U-bends, the thermal behavior of the edge cells, as well as the accuracy of the method.

Nomenclature

A	heat exchange surface (m^{-2})
G	mass flux (kg m^{-2} s^{-1})

i	enthalpy (J kg^{-1})
LMTD	log-mean temperature difference (K)
m	mass flow rate (kg s^{-1})
Δp	pressure drop (Pa)
Q	heat transfer rate (W)
RH	relative humidity
T	temperature (K)
U	overall heat transfer coefficient (W m^{-2} K^{-1})
V	velocity (m s^{-1})
w	air specific humidity (kg^{-1})
x	vapor quality

Greek symbols

β	Regression coefficient

Subscripts

A,R	air, refrigerant
i	inner
in	inlet
m	mean value
o	outer
out	outlet
w	wall

Abbreviations

HX	heat exchanger

Author details

Giuseppe Starace[1*], Silvia Macchitella[2] and Gianpiero Colangelo[2]

1 Department of Management, Finance and Technology, Casamassima (Bari) LUM University, Italy

2 Department of Engineering for Innovation, University of Salento, Lecce, Italy

*Address all correspondence to: starace@lum.it

References

[1] Corberán JM, de Fernandez Cordoba P, Ortuno S, Ferri V, Setaro T, Boccardi G. Modelling of tube and fin coils working as evaporator or condenser. In: Proceedings of the 3rd European Thermal Sciences Conference. Heidelberg; 2000

[2] Corberán JM, García M. Modelling of plate finned tube evaporators and condensers working with R134a. International Journal of Refrigeration. 1998;**21**(4):273-284

[3] Jiang H, Aute V, Radermacher R. CoilDesigner: A general-purpose simulation and design tool for air-to-refrigerant heat exchangers. International Journal of Refrigeration. 2006;**29**:601-610

[4] Tarrad AH, Al-Nadawi AK. Modelling of finned-tube using pure and zeotropic blend refrigerants. In: Proceedings of ATINER'S Conference. Athens; 2015

[5] Yun JY, Lee KS. Influence of design parameters on the heat transfer and flow friction characteristics of the heat exchanger with slit fins. International Journal of Heat and Mass Transfer. 2000: 43:2529–2539.

[6] Matos RS, Laursen TA, Vargas JVC, Bejan A. Three-dimensional optimization of staggered finned circular and elliptic tubes in forced convection. International Journal of Thermal Sciences. 2004;**43**:477-487

[7] Joppolo CM, Molinaroli L, Pasini A. Numerical analysis of the influence of circuit arrangement on a fin-and-tube condenser performance. Case Studies in Thermal Engineering. 2015;**6**:136-146

[8] Sim J, Lee H, Jeong JH. Optimal design of variable-path heat exchanger for energy efficiency improvement of air-source heat pump system. Applied Energy. 2021;**290**:116741

[9] Wang F, Zhao R, Ma C, Huang D, Qu Z. Reversely-variable circuitry for finned-tube heat exchanger in air source heat pump to enhance its overall energy performance. International Journal of Refrigeration. 2022;**142**:48-57

[10] Starace G, Fiorentino M, Longo MP, Carluccio E. A hybrid method for the cross flow compact heat exchangers design. Applied Thermal Engineering. 2017;**111**:1129-1142

[11] Carluccio E, Starace G, Ficarella A, Laforgia D. Numerical analysis of a cross-flow compact heat exchanger for vehicle applications. Applied Thermal Engineering. 2005;**25**(13):1995-2013

[12] Fiorentino M, Starace G. The design of countercurrent evaporative condensers with the hybrid method. Applied Thermal Engineering. 2018;**130**: 889-898

[13] Starace G, Fiorentino M, Meleleo B, Risolo C. The hybrid method applied to the plate-finned tube evaporator geometry. International Journal of Refrigeration. 2018;**88**:67-77

[14] Starace G, Macchitella S, Fiorentino M, Colangelo G. Influence of circuit arrangement on evaporator performance using the hybrid method. In: Proceedings of 6th IIR Conference on Thermophysical Properties and Transfer Processes of Refrigerants. Vicenza; 2021

[15] Starace G, Macchitella S, Colangelo G. Improvements to the hybrid method applied to the design of plate-finned tube evaporators. In:

Proceedings of 77° ATI Conference. Bari; 2022

[16] Starace G, Macchitella S, Colangelo G. The hybrid method for the plate-finned tube evaporator design process. In: Proceedings of 76° ATI Conference. Rome; 2021

[17] Wang CC, Lin YT, Lee CJ. An airside correlation for plain fin-and-tube heat exchangers in wet conditions. International Journal of Heat and Mass Transfer. 2000;**43**(10):1869-1872

[18] Bourabaa A, Saighi M, Belal I. The influence of the inlet conditions on the air side heat transfer performance of plain finned evaporator. The International Journal for Computation and Mathematics in Electrical and Electronic Engineering. 2011;**5**(11):1667-1670

[19] Thulukkanam K. Heat Exchanger Design Handbook. Boca Raton, FL, USA: CRC Press; 2013

[20] Kays WM, London AL. Compact Heat Exchanger. 3rd ed. New York: McGraw-Hill; 1984

[21] Liang SY, Wong TN. Experimental validation of model predictions on evaporator coils with an emphasis on fin efficiency. International Journal of Thermal Sciences. 2010;**49**:187-195

[22] Ma X, Dinga G, Zhanga Y, Wang K. Airside heat transfer and friction characteristics for enhanced fin-and-tube heat exchanger with hydrophilic coating under wet conditions. International Journal of Refrigeration. 2007;**30**(7):1153-1167

[23] Stephan K, Abdelsalam M. Heat transfer correlations for natural convection boiling. International Journal of Heat and Mass Transfer. 1980;**23**(1): 73-87

[24] Choi TY, Kim YJ, Kim MS, Ro ST. Evaporation heat transfer of R-32, R-134a, R-32/134a and R-32/125/134a inside a horizontal smooth tube. International Journal of Heat and Mass Transfer. 2000;**43**(19):3651-3660

[25] Pierre B. Flow resistance with boiling refrigerants – Part 1. ASHRAE Journal. 1964;**6**(9):58-65

[26] Webb RL, Kim NH. Principles of Enhanced Heat Transfer. 2nd ed. New York: Taylor & Francis Inc; 2005

Chapter 3

Hydrogen System of Autonomous Power Supply of Low Power

Mikhail Luchko

Abstract

In the traditional centralized power supply systems, no alternative power source can be used if the main distribution network is damaged by a natural disaster, which makes traditional power systems fragile. In this article, an analysis of autonomous power supply systems has been carried out and an autonomous hybrid power supply installation based on a hydrogen power plant with a capacity of 5 kW has been developed.

Keywords: system, hydrogen, energy, generator, hydrolysis

1. Introduction

People spend most of their day in inside environments; therefore, those environments must be in good condition for these users. In 2019, the building sector accounted for 35% of the world's energy consumption and 39% of gas emissions, the highest ever recorded [1].

With the increasing number of people living in cities, the demand for energy in homes is constantly increasing; thus, smart buildings were created to provide users with better comfort conditions.

In order to increase the efficiency of traditional sources of electricity, such as solar and wind power plants, as well as ensuring reliable power supply of the reserve for consumers, reducing the environmental burden on nature requires very effective means of energy storage and production.

"Increasing the capacity of systems from electrochemical batteries is associated with high costs, they become unreliable after prolonged operation, are sensitive to temperature and are hazardous to the environment during disposal." Also, for a small power generation system, the overall characteristics of energy storage devices become critical. Metal hydride storage technology is a good alternative to other storage systems. "The energy reserve time when using metal hydride storage technology based on hydrogen fuel cells depends only on the amount of stored hydrogen." The hydrogen system does not have a problem of recharging and self-discharge, which guarantees the stability of energy parameters and simplifies the operation [2].

The purpose of the dissertation is to analyze possible options for autonomous power supply systems. In the course of the study, the following tasks were performed:

- An analysis of autonomous power supply systems has been carried out.

- A power supply system for a remote settlement based on a hydrogen system has been developed. Object of research—power plants of backup power supply Subject of research—autonomous hybrid installation energy supply on the basis of a hydrogen power plant with a capacity of 5 kW.

Scientific novelty of the work:

- the required number of elements of the power system for sustainable and uninterrupted power supply to the consumer;

- it has been established that the developed autonomous power supply system based on a hydrogen system is economically feasible in the absence of the possibility of connecting to a centralized power supply system with a service life of more than 20years.

Practical significance of the work:

- the result of the work allows us to establish prospects for the development of the topic of the use of small-scale energy based on a hydrogen power supply system.

Approbation of the work:

The main provisions of the work were presented in the form of a report at student scientific and technical conferences.

Publications. Two printed works have been published on the topic of the dissertation.

Structure and scope of work. The dissertation consists of an introduction, three chapters, and conclusions. In the course of the study, a comparative analysis of classical power supply systems and a low-power hydrogen system up to 5 kW was carried out. A system based on an autonomous hydrogen power supply system was proposed.

Analytical review of information sources on the topic of the work "The current state of global energy, based mainly on traditional hydrocarbon sources of raw materials, is characterized even by optimistic estimates as precrisis." And it is not only the obvious depletion of these resources, but also the increasing danger of global disasters due to environmental pollution. Of course, nuclear energy, taking into account the available uranium reserves in thorium, will play an increasingly important role in the global economy. It is assumed that in the distant future, thermonuclear energy will take a decisive place in large-scale energy. However, it is already clear that serious energy and environmental problems will overtake the world before the first thermonuclear power plant comes into operation. In addition, even if we assume that nuclear and thermonuclear energy will be able to produce the necessary amount of electricity, it remains unclear how the energy obtained can ensure, for example, the functioning of transport or the vital activity of remote areas [3].

Today, transport uses about half of the world's the volume of consumption of petroleum products, and in the USA—up to 65%. At the same time, the exhaust of internal combustion engines contains about 45 toxic substances, including carcinogens. Therefore, the search for alternative renewable and environmentally friendly sources capable of providing humanity with energy for the next hundreds of years is one of the undoubted priorities of modern science. This search shows that one of the most likely substitutes for organic fuels of energy carriers for transport and energy in general is hydrogen. Hydrogen is suitable for all types of heat engines: reciprocating, turbine, piston-turbine, Stirling engines, etc. At the same time, hydrogen as a fuel has a high energy content per unit mass—120.7 MJ/kg, which is higher than that of any organic fuel.

"The use of hydrogen for energy production leads to a sharp decrease in environmental pollution."

During the combustion of hydrogen in oxygen, toxic exhausts are completely absent, since the reaction product is water, and during combustion in air, pollution is much lower than when using gasoline. It is very important that hydrogen can be used for direct conversion of chemical energy into electrical energy. Such a transformation occurs in an electrochemical generator (fuel cell) when hydrogen is combined with oxygen on one of the electrodes, and harmful emissions are practically absent. "The efficiency of a fuel cell can reach very high values—from 40 to 70%, and it depends relatively little on the installed capacity and load," (recall that the efficiency of thermal machines such as internal combustion engine, diesel, does not exceed 40%). It is the progress in the development of fuel cells (TE) with high efficiency that inspires confidence in the prospects of using hydrogen as a fuel when creating autonomous mobile and stationary energy sources. Such sources can be widely used in transport, including in cars with so-called "hybrid" engines [4] (a conventional engine plus an electric motor on a fuel cell).

"Cars with TE are especially promising for use in urban environments. Another booming TE market is associated with the need to increase the duration of continuous operation of small-sized electronic devices (cell phones, portable personal computers, etc.) and replace conventional batteries and accumulators in them with more energy-intensive power sources" [5]. The successes achieved in the development of fuel and energy sources, the rise in the value of traditional energy carriers (especially oil), political instability in oil-exporting countries, environmental problems—all this has led to an awareness at the government level of the need for accelerated development of research and technology in the field of hydrogen energy. In this regard, the decision of US President George W. Bush to include hydrogen energy among the national priorities is characteristic. The US Congress has decided on financing an amount of $ 1.3 billion. USA works on fuel cells for cars. Japan supports the development of technologies based on hydrogen and TE through a 28-year program (1993–2020) with a total budget of 2.4 billion euro. Financing of research on TE in Europe is approximately 1/3 of the financing in the USA and 1/4 of the financing in Japan. At the same time, in recent years, EU countries have been actively developing a strategy to consolidate the efforts of governments and large international companies in the development of hydrogen technologies and fuel cells. It should be noted that large nonstate companies, mainly automotive ones, also invest large funds in the development of hydrogen technologies.

2. Experimental study of hydrogen production using solar energy

Hydrogen is a sustainable fuel option and one of the possible solutions to environmental problems. In this study, hydrogen is obtained using a hydrogen generator with a proton-exchange membrane (POM) of an electrolyzer. A pilot study is being conducted at the Renewable Energy Development Center in Algeria.

The experimental device contains: a photovoltaic module, a POM electrolyzer, a gasometer, and devices for measuring the characteristic POM electrolyzer, as well as two pyranometers and a diffuser. This system allowed, on the one hand, to measure and analyze the characteristics of the POM electrolyzer at different pressures (Ratm and p = 3 bar), and on the other hand, allowed to study the volume of hydrogen in various sources of electricity (generator, photovoltaic module, and fluorescent lamp). The efficiency for each case was calculated and compared. This article presents the values of the change in the rate of hydrogen flow, depending on the time of day the experiment was conducted in which was in August during the daytime.

Hydrogen is considered as the energy carrier of the future. It can be obtained in various ways, including using solar energy, as well as using solar thermal energy.

Solar energy can be used in an electrolyzer to decompose distilled water into hydrogen and oxygen. Autonomous electrolyzer systems are used for the production of hydrogen fuel. This system consists of a photovoltaic module that supplies electricity through an electrolyzer system, as shown in the major projects in USA and FRG.

Water electrolysis is considered one of the key technologies for hydrogen production because it is compatible with existing and future power generation technologies and many renewable technologies (solar, biomass, hydro, wind, tidal, etc.). Most water electrolysis technologies currently available on the market use acidic or alkaline electrolyte systems for hydrogen production. Typical efficiencies are shown in the 55–74% range, with most commercial systems having efficiencies of less than 65%. Current densities are typically around [6] 0.3–0.4 A/cm, and there are technical difficulties in maintaining the electrolyte balance, as well as in maintaining hydrogen and oxygen. Currently, the technology of electrolysis of water based on a polymer electrolyte membrane is under development. The POM electrolysis system can quickly respond to changes in energy consumption and, therefore, can be easily integrated with renewable energy systems. The POM operates at relatively low temperatures, usually at 80°C or below, and usually consists of numerous cells stacked in series. Guillaume Doucet and others studied the general characteristics of the integrated and automated hydrogen power plant (APU).

The system consists of a 0.46 W photovoltaic module (RV), and the 0.64 W electrolyzer that includes a proton exchange membrane, an alternating voltage generator, a 1000 W fluorescent lamp, two vessels with a capacity of 250 ml, and a device that removes the characteristics of the POM electrolyzer, as well as two pyranometers [7].

The main part of the POM block is a membrane electrode block. A layer of catalyst material was applied to both sides of a thin proton-conducting membrane (POM = proton exchange membrane). These two layers form the anode and cathode of the electrochemical cell. As we see in **Figure 1**, oxygen gas, electrons, and H+ ions are formed on the anode side. H+ ions pass through the membrane to the cathode and form hydrogen gas with electrons flowing through an external conductive circuit. Thus, electrical energy is converted into chemical energy, and stored in the form of hydrogen and oxygen.

The electrolyzer produces hydrogen and oxygen in a ratio of 2:1 (**Figure 1**), the volume of hydrogen released is measured as a function of time, that is, the start of the countdown occurs when the water in the gasometer (H2) passes the lower sign. Next, the voltage and current I are measured during electrolysis [6].

The experience done should first of all show how the characteristics of the POM electrolyzer change at different pressures.

At the experimental facility, we determine the volume of hydrogen produced at different energy sources. The first way is solar energy. The second method consists

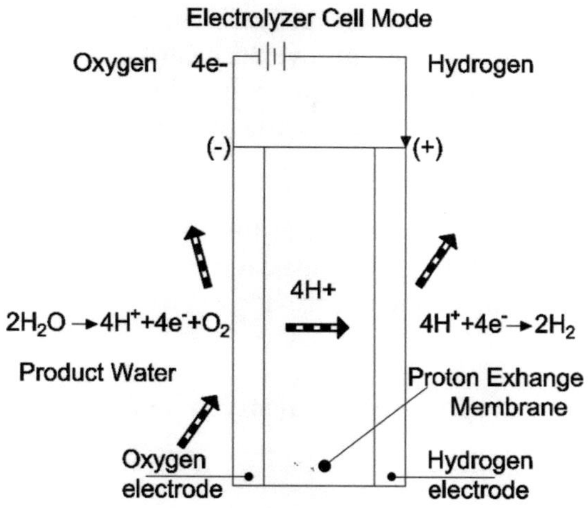

Figure 1.
Functional principle of the POM electrolyzer.

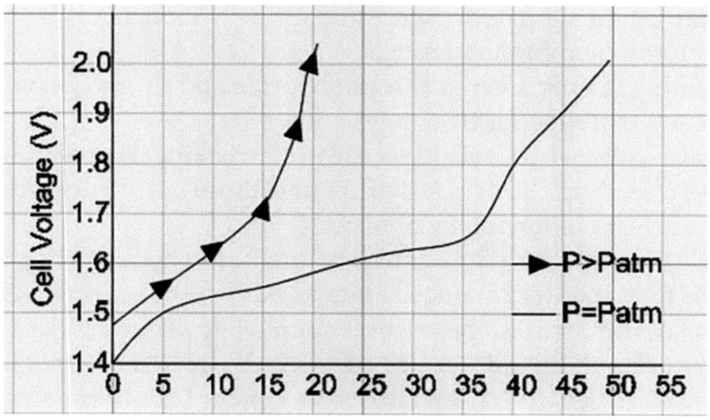

Figure 2.
Graph shows that the pressure has a significant impact on the performance.

of simulating solar energy using a lamp. The third method is an electrolyzer POM voltage generator.

The results are presented in the following paragraphs. First, the POM electrolyzer is operated at two different pressures (Patm and P = 3 bar). Then the volume of hydrogen was produced with various sources of electricity (generator, photovoltaic module, and daylight lamp). Finally, we calculated the efficiency for each case.

The performance of the POM electrolyzer for two different pressures is shown in **Figure 2**.

The graph shows that the pressure has a significant impact on the performance of the electrolyzer POM. The effect of increasing oxygen pressure from three bars leads to an increase in voltage of 0.06 V and current. These results show that an increase in oxygen pressure leads to a significant improvement in polarization at the cathode.

The conclusions obtained:

- Hydrogen is a clean source of energy. In addition, the energy required to create hydrogen requires less than that can be derived from it.

- Reducing dependence on fossil and nuclear energy would free the country from costly foreign involvement and improve the health and welfare of its citizens. This would be accomplished by significantly reducing greenhouse gas emissions and other toxins associated with fossil and nuclear energy sources in the air, land, and water [6].

- Solar energy can serve as a power source for the electrolyzer to produce hydrogen. Hydrogen will be stored in storage devices to start the thermal power plant when solar energy is not enough to provide electricity to the consumer.

3. Development of hydrogen storage systems based on complex metal hydrides

This review describes the latest research in the development of a tank based on complex metal hydrides for thermolysis and hydrolysis. Commercial applications using complex metal hydrides are limited, especially for thermolysis-based systems, where only demonstration projects have been carried out so far. Hydrolysis-based systems find their application in the space and military industries due to their compatibility with proton membrane fuel cells.

Tanks containing mainly sodium aluminohydride and several examples with nitrides have been developed for thermolysis.

For hydrolysis, sodium borohydride is the preferred material, while ammonia has proven to be less effective. The disposal of spent sodium borohydride fuel remains an important part of their commercial viability.

Over the past 15 years, complex aluminum and boron hydrides have been investigated as possible materials for hydrogen storage. Although the composition of these materials is similar, the chemical behavior is completely different. An example of some complex aluminum hydrides are NaAlH4, CAlH4, NA1H6, etc. These hydrides can be decomposed at elevated temperatures and under technically significant conditions using catalysts, and repeated dehydrogenation can also be carried out. However, the decomposition temperature of complex boron hydrides (LIH4, NVH4) is often much higher, and reversibility cannot be observed under the conditions used for complex aluminum hydrides. Consequently, complex aluminum hydrides can be used for technical applications where repeated dehydrogenation of the material for hydrogen storage is an important condition [8]. And complex boron hydrides are preferred for cartridge systems that release hydrogen in a hydrolysis reaction at ambient temperature. These different properties lead to completely different technical requirements. This makes the development of storage systems based on aluminum or borohydride compounds a difficult task.

Thermolysis requires heat input and care must be taken when designing a storage tank to distribute heat efficiently. Hydrolysis, on the other hand, requires not only the efficient mixing of complex hydrides and water but also the separation of the resulting hydrogen gas from a suspension composed of decomposition products and water. While pyrolysis tank systems are being developed in demonstration projects, hydrolysis storage systems are already being applied in real-world settings [8].

3.1 Autonomous hybrid power supply unit based on a hydrogen power plant

The structure of the system under consideration includes

- hydrogen generator;
- fuel cell module;
- inverter (voltage converter to ~220 volts);
- metal hydride storage.

1-external source; 2- voltage stabilizer; 3-hydrogen generator; 3- desiccant; 5-compressor; 6- check valve; 7- metal hydride storage; 8-reducer; 9-fuel cell; 10-ECU; 11-supply valve distilled water; 12- H20; 13- consumer; 14- A-ammeter; 15 -DH1-voltage sensor; 16- K1, K2- circuit key; 17 DU- level sensor; 18 UPD- hydrogen supply control; 19 DD- pressure sensor (**Figure 3**).

The installation works as follows: the first circuit of the circuit is an external source, which is a power line. Its work consists in the energy supply of the consumer. The second circuit of the scheme is a hydrogen power plant with a nominal capacity of 5 kWh, which provides the consumer with electricity in the absence of electricity from an external source.

The Dantherm Power DBX5000 fuel cell module is shown in **Figure 4**—with a power of 5 kW. The average price of a fuel cell will be 550 € (**Figure 5**).

The technical characteristics of the generator are given in **Table 1**.

The operation of the hydrogen generator involves water costs, so we will give the cost calculations that are given in **Table 2**.

Metal hydride storage of hydrogenBL-3 (**Figure 6**). The average price is 500 EUR. We need two drives. Therefore, the price will be 1000 EUR (**Table 3**).

The price of owning the installation for 10 years will be 8125 EUR.

- Hydrogen is a clean source of energy. In addition, the energy required to create hydrogen requires less than can be derived from it.
- Reducing dependence on fossil and nuclear energy will free the country from costly foreign involvement and improve the health and well-being of its

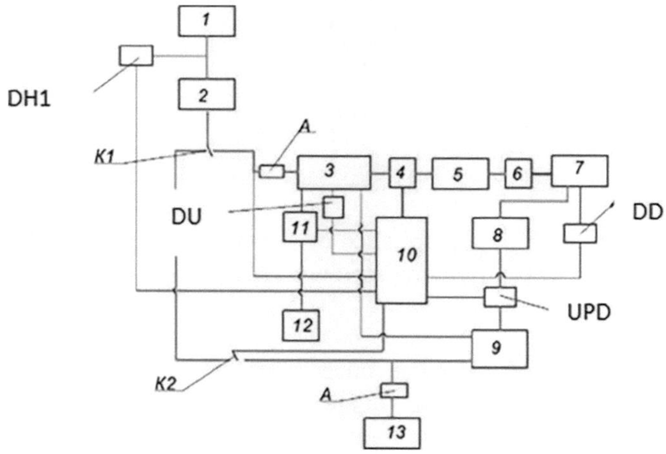

Figure 3.
Diagram of an autonomous hybrid power supply unit based on a hydrogen power plant.

Efficiency	5000 watts
Output voltage VDC	47-53 VDC (48 VDC)
Input voltage VAC	90-264 / 50-60 Hz
Pressure at the inlet to the valve block	5 bar
Duration of operation on one cylinder based on the use of a 40 liter cylinder	~ 150 min
Ambient temperature requirements	-20°C до +50°C
Humidity requirements	5-95%
Height range	up to 3000m above sea level

Figure 4.
Dantherm power DBX5000.

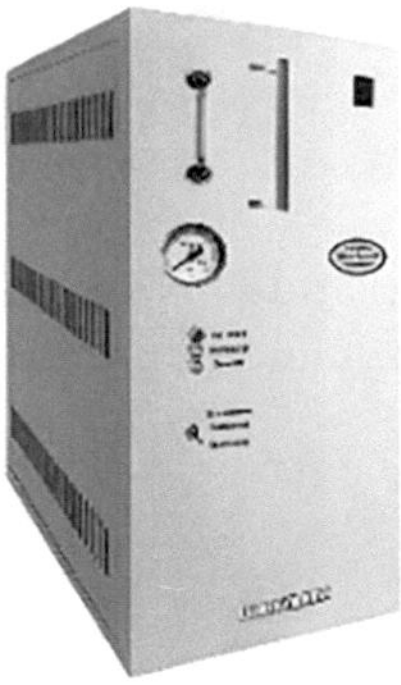

Figure 5.
Hydrogen generator.

citizens. This would be accomplished by significantly reducing greenhouse gas emissions and other toxins associated with fossil energy sources in the air, land, and water [6].

- Solar energy can serve as a power source for the electrolyzer to produce hydrogen. Hydrogen will be stored in storage devices to start the fuel cell when solar energy is not enough to provide electricity to the consumer.

Purity of hydrogen in terms of dry gas %	99,995
Concentration of water vapor at 2OC and 1 atm, no more, rrsh	5
Total hydrogen capacity reduced to normal conditions, l/h	16
The range of the specified output pressure of hydrogen, atm	1.5-6.5
Stability of hydrogen output pressure, atm	± 0.02
Operating mode setting time, no more than, min	30
Volume of distilled water to be poured, liters	1.0

Table 1.
The technical characteristics of the generator.

	Working/Hours	Water/Liters	Price/EUR
Day	10	0.1	0.01
Month	2100	21	2.1
Year	25,200	252	25.2

Table 2.
Calculation of water costs for the year of operation.

Figure 6.
Metal hydride storage of hydrogen BL-3.

Volume, liters	25
Pressure, atm	5
Length, millimeters	147
Weight, grams	307

Table 3.
Technical characteristics of a metal hydride storage device.

4. Conclusions

A diversity of research on strategies to promote thermal comfort and energy efficiency in buildings for sustainability was examined in addition to the main challenges and barriers encountered. Key aspects of thermal conditions that affect energy efficiency were identified in addition to building types and climates, new technologies, and discoveries to provide thermal comfort and reduce energy consumption.

- An autonomous energy supply system based on a hydrogen system can be considered under modern solutions, since it corresponds to current global trends in the field of power supply systems.
- The introduction of a hydrogen fuel cell system will implement environmentally friendly solutions for backup power supply systems.
- In terms of functional characteristics, fuel cell-based SRES are significantly superior to existing systems, which are confirmed in implemented projects all over the world.
- By cost indicators for averaged (typical) source, data backup power systems for fuel and energy sources and the initial cost is slightly higher in terms of cost of ownership by 10–30% as compared to the existing ones.

Author details

Mikhail Luchko
Angrsk State Technical University, Angarsk, Russia

*Address all correspondence to: luchkomw@gmail.com

References

[1] UNEP—United Nations Environment Programme. 2020 Global Status Report for Buildings and Construction: Towards a Zero-emissions, Efficient and Resilient Buildings and Construction Sector—Executive Summary. Available from: https://wedocs.unep.org/20.500.11822/34572. [Accessed: July 25, 2022]

[2] Nassar A, Nassar E. The design of a low-cost device for the production of hydrogen. Science. 2014;**13**(25):158-165

[3] Carrión N, Murillo M, Rodríguez H, José Chirinos, Diaz D. Study of the fundamental plasma parameters by HG ICP-OES with a dual hydride generation system. Science. 2011;**6**(3):61-68

[4] Adinarayana G, Ashok Kumar C, Ramakrishna M. Fabrication of hybrid petroelectric vehicle. Science. 2014;**4** (10 (Part - 6)):142-144

[5] Bendaikha W, Larbi S, Mahmah B, Belhamel M. Experimental study of the production of solar hydrogen in Algeria. Science. 2010;**2**(1):192-202

[6] Lanez T. University of El Oued. JFAS. 2020:1905-1982. Available from: https://www.jfas.info/

[7] Morten B. Ley , Mariem Meggouh, Romain Moury, Kateryna Peinecke and Michael Felderhoff. Development of hydrogen storage tank systems based on complex metal hydrides. Science. 2015;**8**(9):5891-5921

[8] SciProfiles is a Platform Maintained by MDPI. 2023. Available from: https://sciprofiles.com/

Chapter 4

The Performance and Characteristics of the Cooling System of Processors in Data Centres

Raksha Manel Shenoy

Abstract

In this contemporary world, technology is advancing at a rapid pace. In recent decades, information technology has grown exponentially. Improving data centre energy efficiency is an urgent issue with remarkable economic and environmental impacts. Enormous heat generation is fuelled by data centres. The distribution of heat is a crucial parameter that affects data centre cooling and energy consumption. Heat reduction is the crux of environmentally sustainable computing. Typically, data centres lack in effective sensing systems to monitor heat distribution at a large scale. In this paper, I aim to use sensor networks as a dense instrumentation technology to comprehend and control cooling in data centres. I present the Aquasar project, as a case study and explore environmental challenges in developing sensor networks in data centres. I also explore the effective way of dealing with the excess heat from multiple servers in the data centres. And the upper hand of water cooling over air cooling.

Keywords: water-cooling, air cooling, data centre, processor, Aquasar

1. Introduction

Green computing is the survey and methodology of naturally renewable computing. It encompasses planning, fabricating, utilizing and getting rid of computers, servers, and related subsystems such as screens, printers, storage drives and organizing the communications frameworks productively and viably with negligible or no effect on the environment, as in [1]. Data centres, which have been criticized for their extraordinarily high energy demand, are a primary focus for exponents of green computing. E-waste refers to the waste, unusable electronic and electrical goods. All electronic items have generally elements such as lead, beryllium, etc., which, if untreated, may cause harmful effects. Thus e-waste treatment is a very important part of environmental management. To make matters worse, e-waste generation has been growing exponentially over the past few years. The main culprits are the data centres. Data centres are places with many computers serving as servers or storage devices. According to StEP (Solving the E-waste problem) initiative, data centres are responsible for 60% of the e-waste coming into the world. Based on current trends, E-waste will grow from 48.9 million metric tonnes in 2013 to 65.4 million tonnes in 2017; additional information can be found in section [2]. Their e-waste generation has alarmed environmental enthusiasts, and they need to be given attention.

The e-waste is generated mainly because either they go for up gradation or complete renewal. The main e-waste generated is that of processors. Since they have the least lifetime and the most amount of work, they get wearer easily and thus require a lot of attention. Processors have a property of over-clocking, which is used to get higher speeds at times, which generates a lot of heat, since the processor goes beyond its capacities and provides a higher speed. This heat generation almost halves the processor life, as given in section [3]. The main reason why processors are thrown is their degradation due to heat and dust. The processor lifetime is about 3 years; due to cooling, I can upgrade it to about 6 years. Thus, the data centre, which was throwing away processors every 3 years, will now throw away the same processors after 6 years, thus is reducing the e-waste by a direct factor of half. In this way, it will help to "reduce" e-waste.

The method discussed here, as the case study requires an extensive use of copper, therefore giving a chance for the e-wastes containing copper to get used up again. Thus, I am "reusing" the e-waste components.

2. Air cooling

Figure 1 is a pictorial representation of the cross-section of a data centre room which is also called server collocation (*colo* for short). Racks are installed on the elevated floor in aisles. Computer room air conditioning (CRAC) systems supply cool air to the subfloor. To make cool air available to servers, some floor tiles are holed, serving as vents. The aisles, along with these vents, are called cold aisles. Usually, cool air is drawn from the front, and hot exhaust air is blown to the back – hot aisles. Servers are arranged face to face adjacent to the cold aisles for the efficient use the cool air. As shown in **Figure 1**, cool air and hot air are ultimately mixed near the ceiling and is drawn into the CRAC where this mixed air exchanges heat with chilled water. For regulating the temperature to a set point, as in [4] chilled water valve opening is controlled, and a temperature sensor is located at the intake of the CRAC. Air cooling was the technique used initially in the data centres.

Cooling was done by using air in most of the data centres for practical reasons. Air is abundant, it normally poses no threat to humans or equipment, it is a bad

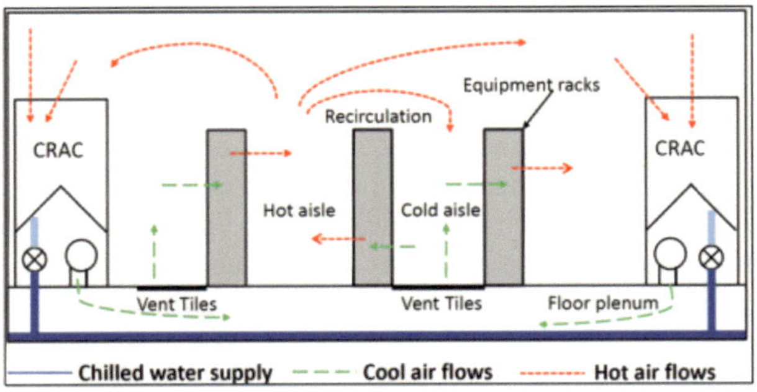

Figure 1.
Cross-Section of colo as in [4].

conductor of electricity, it's easy to propagate and it's free of cost. But air also falls short on several counts; for example, the thermally efficiency of air is less (i.e., it does not hold much heat relative to liquids), thus cooling of high-density implementations is not feasible, details of which are provided in [5]. And moreover, the cold air may just pass between the aisles instead of within the server. That is how air cooling precisely fails. In order to increase efficiency, most of the companies are using water (or some other liquids with similar properties as of water) and provide more-targeted cooling to achieve high-power computing and similar conditions. Water, mainly because it is cheap, readily available and has the highest specific heat in comparison with other liquids.

3. Water cooling

The power loadings are incorporated in the novel designs of data centres. Other cooling methods, like phase change liquid systems or single-phase forced liquid flow, are now being used [6, 7]. Using liquid cooling systems can result in significant reductions in the overall cooling energy need, as mentioned by Greenberg et al. [8].

The requirement for both Computer Room Air Conditioning (CRAC) units and chiller plants can be decreased or, in some circumstances, potentially eliminated using ambient free-air cooling and the use of water/air side economizers, resulting in significant energy savings. According to Brunschwiler et al. [9], water-cooled systems have modest temperature fluctuations throughout the system because of more effective heat transmission. The microprocessor junction temperature can be kept far below 85°C with a water flow rate of 0.7 L/min and an input temperature of 60°C, and if the chip is allowed to get close to 85°C, the intake water temperature can get as high as 75°C [9]. The waste heat produced by these higher temperatures for liquid cooling systems will be of greater quality, and energy absorption using the direct discharge liquid stream will be simpler. Chillers may not be necessary with the higher-temperature coolant, which lowers the energy consumption for the system. This technique places the cold plates as near as feasible to the parts that generate heat [9]. These systems' thermal resistance, which can remove more than 200 W/cm^2 of heat, is less than 20% of air-cooling systems' thermal resistance [9].

The benefits of water-cooled systems over air-cooled ones are illustrated by a thorough comparison study comparing the energy efficiency of water-cooled and air-cooled high-density servers, carried out by Ellsworth and Iyengar [10]. According to their research, employing water cooling can boost processor performance by 33%. All energy-dissipating components, such as the CPU, memory, power conversion, and I/O circuits, are cooled by aluminum/copper tube cold plates in another water-cooling study by Ellsworth et al. [11]. In this instance, condensation was prevented by setting the cold plate supply temperature at least 7°C above the dew point. The effectiveness of a water-cooling system for an electronic module containing a 150 W dual-core CPU and an 8 W memory chip was evaluated by Campbell and Tuma [12]. According to their calculations, the maximum temperature of water entering the module must be higher than 28°C in order to maintain junction temperature at 65°C with a water flow rate of 0.95 L/min. In a dual loop chiller-less data centre that IBM recently developed and built, the cooling energy need was decreased from the industry standard of 45% of the total energy consumed by the data centre to just 3.5% [13–15]. 38 "warm water" cooled servers are housed in a single rack in this test-scale data centre. Each server has cold plates for the processors and memory modules, and recirculated air that has been pre-cooled by the water entering the servers is used to cool the remaining components.

The requirement for a CRAC unit is essentially nonexistent because the water loop removes most of the server heat dissipation. The cooling water from the rack is circulated internally in a loop with heat transfer to an external coolant loop, like a water-glycol solution. Using a dry cooler, heat is transferred from the closed external loop to the surrounding air without the use of make-up water, as is the case with wet cooling tower methods. The water flow rate varied from 4 to 8 GPM, and the rack power ranged from 13.4 to 14.5 kW [13–15]. To balance maximizing waste heat recovery with maintaining the thermal stability of the chip, Sharma et al. [16] found that the ideal water inlet temperature for cooling a microprocessor chip should be between 40 and 47.5°C at an ideal flow rate of 1 L/min.

4. Aquasar project

As an attempt to have efficient cooling methods, IBM developed this project, mainly for the benefit of supercomputers and data centres. IBM and the Swiss Federal Institute of Technology are working on a water-cooled supercomputer whose excess heat will be utilized to heat the university's buildings. This project is IBM patented and unique since it uses warm water instead of cold water for cooling mechanism. Water, which is about 4000 times more efficient as a coolant than air, is made to enter the system at 60°C, keeping the chips in the system at operating temperatures below their maximum of 85°C. A high-grade heat (which in this case will be about 65°C as in [17]) is obtained as an output due to high input temperature.

The system makes use of jet impingement cooling, which involves direct contact of water with the back of the chip via microchannels in the heat sink. "This method neither causes the overhead in thermal resistance of a base plate, nor the overhead and reliability problem of thermal interface materials, and thus is responsible for removing highest-power densities," according to one paper published by the scientists at IBM as in [17]. **Figure 2** depicts the connection between the pipelines from the individual blades and the server rack's water-pipe network, which is further connected to the main water transportation network. Aquasar will need about 10 liters of water for cooling, pumped at some 30 liters per minute as described in section [17].

Figure 2.
Water flows along copper pipes in a blade server used in the Aquasar supercomputer as in [17].

In a closed-circuit cooling system, chips are used to heat the water, which is subsequently cooled to the required temperature as it passes through a passive heat exchanger, distributing the discarded heat directly to be used later. However, water and electrical components are isolated from each other.

My work on this project was to focus on the performance and characteristics of the cooling system which will be measured with an extensive system of sensors, to optimize it further. I propose the following sensor algorithms.

Sensor Algorithms

Algorithm for temperature sensing in data centres

STEP 1: start

STEP 2: repeat STEP 3 until temperature belongs to optimal temperature range (depends on the processor used)

STEP 3: do nothing

[End of STEP 2 loop]

STEP 4: signal the OS

STEP 5: repeat STEP 6 until temperature > optimal temperature or temperature < optimal temperature

STEP 6: continue signaling the OS

[End of STEP 5 loop]

STEP 7: go to STEP 2

STEP 8: END

Note: [temperature < optimal temperature is checked to prevent undercooling of processors]

Algorithm for pressure sensing in data centres

STEP 1: start

STEP 2: repeat STEP 3 until pressure belongs to optimal pressure range exerted by the liquid flowing in the copper pipes.

STEP 3: do nothing

[End of STEP 2 loop]

STEP 4: signal the OS

STEP 5: repeat STEP 6 until pressure > optimal pressure or pressure < optimal pressure

STEP 6: continue signaling the OS.

[End of step 5 loop]

STEP 7: go to STEP 2

STEP 8: END

5. Sensor deployment

The sensors used are wireless as it enables real-time decisions to operate a data centre at peak efficiency. This real-time capability ensures streaming data packets to gateway devices that relay the data to software for visual display and interpreted for operational adjustments as in [18]. Temperature sensor is placed above the server cabinet (see **Figure 3**). Pressure sensor is placed between the pipes used to circulate the coolant to monitor the pressure of the liquid coolant flowing in the pipes.

When the temperature of the processor exceeds the allowed temperature (50–95°F) range, temperature sensor detects this and signals the operating system (OS), which in turn signals the motor to increase the quantity of water flowing in the pipes.

Overheating of the processor due to over-clocking is prevented.

Figure 3.
Stacked server racks with sensors as in [18].

When the temperature of the processor falls below the allowed temperature (below 50°F), the temperature sensor detects this and signals the operating system (OS), which in turn signals the motor to decrease the quantity of water flowing in the pipes. Overcooling of the processor is prevented.

Similarly, if the pressure falls below the required pressure (depending on the radius of the pipe, the nature of the liquid and several other factors), the pressure sensor detects this and signals the operating system, which in turn signals the motor to increase the water flow in the pipes.

If the pressure exceeds the required pressure, the pressure sensor detects this and signals the OS, which in turn signals the motor to decrease the water flow.

If the temperature and pressure sensors signal the motor directly, input to the motor is confusing. This results in error. There are four possible cases.

- Temperature and pressure sensors signal the motor to increase the water flow.
- Temperature and pressure sensors signal the motor to decrease the water flow.
- Temperature sensor signals to decrease the water flow and pressure sensor signals to increase the water flow and vice versa.

Therefore, sensors should signal the occurrence of these events to the OS and not the motor. The whole point of an event is to notify a listener that something has happened to a part of the User Interface (UI). An event incorporates all the data that a listener ought to find out what took place and in which component it happened. The data should be whole and complete so that the listener can find out what happened precisely and respond accordingly [19]. OS controls the motor states via listeners, namely temperature and pressure. It is possible to create your own event listener method with appropriate commands to handle the event.

If the water requirement cannot be met as the processor executes too many processes, OS must kill some processes. Different algorithms can be used to choose the victim process.

6. Thermal-responsive server rearrangement to reduce thermal stress

An increment in inlet air temperature may decrease the rate of heat emanation from air-cooled servers. In this way, thermal stress is established in these servers. As a result, the hotspot locales of thermal stress originate due to the ineffectively cooled active servers that begin conducting heat to the adjacent servers. Thus, failure in the hardware of these servers may result in performance and financial loss consuming higher energy for the cooling mechanism. Therefore, thermal profiling needs to be done.

This segment illustrates the model to reduce the thermal stress by relocating the thermal-aware server. All servers likely to initiate hotspots or have been a part of them are considered to make a thermal profile of all the data centre servers concerning inlet temperature. To serve this purpose, inlet temperature effect (ITE) thermal benchmark test can be used. The result of this test points to the change in a server's outlet temperature in accordance with changes in inlet temperature at full and zero CPU utilization levels. (Since, CPU is the most power-consuming and the most heat-dissipating hardware component of any computer system, it makes up most of the phrase server utilization.) The values of *Ti*max inlet and $T^{i}{}_{\text{outlet}}$ (increased) can be concluded from ITE test. Homogenous servers have the same $T^{i}{}_{\max}$ inlet. However, *Ti*outlet (increased) is influenced by the location of temperature monitoring sensors which can be verified with multiple tests with different sensor locations as described in [20].

7. Thermal state transition

This section demonstrates a finite set of thermal states for a server within a data centre with respect to inlet and outlet temperature, thermal stress and server utilization level. A thermal state of a server*i*can be defined as a tuple ($T^{i}{}_{\text{inlet}}$, $T^{i}{}_{\text{outlet}}$, μ, σ^{i}) denoted by $S^{i}{}_{n}$, where *n* is a whole number ranging from 0 to 3 as per the state transition diagram as shown in **Figure 4**.

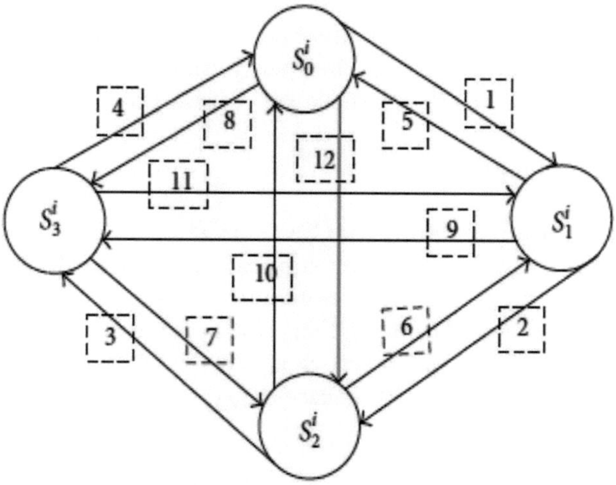

Figure 4.
State diagram of the datacentre as in [20].

As per the assumption, for all the states $T^{i}_{received}<T_{threshold}$, the states S^{i}_{0} and S^{i}_{1} are the desired states with no thermal stress. State S^{i}_{0} is the idle state where the server has no workload μ, and the inlet temperature is close to the set temperature T_{set}. State S^{i}_{1} is an active state of the server where μ is not idle, and the inlet temperature is the same as that of S^{i}_{0}. Both these initial states have outlet temperature below the red line temperature T threshold. There exists a difference in the states S^{i}_{2} and S^{i}_{3}. This is because S^{i}_{2} indicates future thermal stress as hotspots. However, due to a higher inlet temperature than state S^{i}_{2}, state S^{i}_{3} may have thermal stress. Moreover, because the outlet temperature is more than the maximum threshold and the presence of thermal stress as in section [20], state S^{i}_{3} turns out to be a hotspot state.

8. Understanding the economic factors

As the initial hardware investment cost is high for adopting water-cooled data centres, this issue may provoke some companies without going for it, which is between 10 and 30% more than for conventional data centres, according to survey details in [21]. Appropriate copper pipe-based cooling and heat distribution networks to be in place are required by IBM's model. However, the benefits of going green are that computer room air conditioners and chillers are not required in the data centre, which will slash energy costs by up to 50% and dramatically lower the facility's carbon footprint. From a business perspective, the metric that exists and which can give a slightly better idea of what a business measures itself on is the carbon footprint. A carbon footprint measures the total greenhouse gas emissions caused directly and indirectly by a person, organization, event or product, as in [22]. The overall total cost for an air-cooled data centre is greater than that of the ownership cost.

Moreover, buying new processors will cost more than having an efficient cooling system. Thus, it is economically feasible too.

9. Awareness of implications to society at large

The reduction of carbon dioxide emissions will benefit society as a whole. It will be reduced, as said, by 30 tonnes annually. And according to figures in [23], energy consumption will be lowered by 40%. The energy crisis like the one that occurred in since 1972 will be prevented due to the lower consumption of energy.

9.1 Awareness of contemporary issues

The only contemporary issues are that I will have to convince companies to go ahead with these methods. The design must be carefully planned, because any small leakage of water, can damage the whole system.

Another issue can be that of sensors, their working and lifetime. Since they must be working for all 6 years with full-on efficiency. If not, they will reduce the efficiency at large and may cause hazards in case they do not perform their required task.

10. Understanding the ethical and professional issues

Professional challenges will only arise when trying to persuade data centers to accept and use the architecture.

Nearly no ethical problems exist; the only problem is that some individuals might not agree with the procedure because I'm letting water directly touch the CPU.

11. Conclusion

A proper comparison is provided between conventional air-cooling techniques employed in the data centres and modern water-cooling methods. The necessity to employ efficient water-cooling techniques is elaborated. A detailed analysis of sensor algorithms is proposed in this paper. This deployment of sensors further improvises the much-acclaimed Aquasar project. Thus, processor cooling is one of the most important aspects of green computing. The heat captured can be used to warm the building in winter and provide year-round hot water for bathroom and kitchen use. Excess heat can be provided to the swimming pools. Heat can be converted to other forms of energy and used for various purposes.

Author details

Raksha Manel Shenoy
Independent Researcher, Germany

*Address all correspondence to: rakshamanel08shenoy@gmail.com

References

[1] Clark J. A Journal Article from The Data Centre Journal, proposed at The GraphLab Conference. San Francisco: The Data Centre Association; September 2023. Available from: www.datacentrejournal.com/.../liquid-cooling-the-wave-of-the-present/

[2] Murugesan S. Harnessing Green IT: Principles and Practices. Florida, USA: IEEE IT Professional; 2008. pp. 24-33

[3] Rien Dijkstra. The Fundamental Problem of it and data centre e-waste [online] 11 January 2014. Available from: http://infrarati.wordpress.com/category/e-waste-2/

[4] Goth et al. Thermal and mechanical analysis and design of the IBM Power 775 water cooled supercomputing central electronics complex. In: 13th IEEE Intersociety Conference. San Diego, USA: Institute of Electrical and Electronics Engineers; 2012. pp. 700-709. DOI: 10.1109/ITHERM.2012.6231496

[5] Patel CD, Bash CE, Sharma R, Beitelmal M, Friedrich R. Smart cooling of data centres. In: Proceedings of International Electronic Packaging Technical Conference and Exhibition. Maui, Hawaii: The American Society of Mechanical Engineers; June 2003

[6] Schmidt RR, Cruz EE, Iyengar MK. Challenges of data center thermal management. IBM Journal of Research and Development. 2005;**49**(4/5):709-723

[7] Joshi Y, Samadiani E. Energy efficient thermal management of data centers via open multi-scale design: A review of research questions and approaches. Journal of Enhanced Heat Transfer. 2011;**18**(1):15-30

[8] Greenberg S, Mill E, Tschudi B, Rumsey P, Myatt B. Best practices for data centers: lessons learned from benchmarking 22 data centers. In: Proceedings of 14th ACEEE Summer Study on Energy Efficiency in Buildings. Pacific Grove (CA, USA): ECEEE Organization; 13-18 August 2006

[9] Brunschwiler T, Smith B, Ruetsche E, Michel B. Toward zero-emission data centres through direct reuse of thermal energy. IEEE Certified. 2009;**53**(3):11:1-11:13. DOI: 10.1147/JRD.2009.5429024

[10] Ellsworth MJ, Iyengar MK. Energy efficiency analyses and comparison of air and water cooled high-performance servers. In: Proceedings of IPACK0 09. San Francisco (CA, USA): CERN; 19-23 July 2009

[11] Ellsworth MJ, Goth GF, Zoodsma RJ, Arvelo A, Campbell LA, Andrel JW. An overview of the IBM power 775 supercomputer water cooling system. In: Proceedings of IPACK0 11. Portland (OR, USA): CERN; 2011

[12] Campbell L, Tuma P. Numerical prediction of the junction-to-fluid thermal resistance of a 2-phase immersion-cooled IBM dual core POWER6 Processor. In: Proceedings of the 28th IEEE SEMI-THERM Symposium. San Jose (CA, USA): Institute of Electrical and Electronics Engineers; 18-22 March 2012. pp. 36-45

[13] David MP, Iyengar M, Parida P, Simons R, Schultz M, Schmidt R, et al. Experimental characterization of an energy efficient chiller-less data center test facility with warm water cooled servers. In: Proceedings of the 28th IEEE SEMI-THERM symposium. San Jose (CA, USA): Institute of Electrical and Electronics Engineers; 18-22 March 2012. pp. 232-237

[14] Parida PR, David M, Iyengar M, Schultz M, Gaynes M, Kamath V, et al. Experimental investigation of water cooled server microprocessors and memory devices in an energy efficient chiller-less data center. In: Proceedings of the 28th IEEE SEMI-THERM Symposium. San Jose (CA, USA): Institute for Electrical and Electronics Engineers; 18-22 March 2012. pp. 224-231

[15] Iyengar M, David M, Parida P, Kamath V, Kochuparambil B, Graybill D. Server liquid cooling with chiller-less data center design to enable energy savings. In: Proceedings of the 28th IEEE SEMI-THERM Symposium. San Jose (CA, USA): Institute of Electrical and Electronics Engineers; 18-22 Mar 2012. pp. 212-223

[16] Sharma CS, Zimmermann S, Tiwari MK, Michel B, Pulikakos D. Optimal thermal operation of liquid-cooled electronic chips. International Journal of Heat and Mass Transfer. 2012;**55**:1957-1969

[17] EDN Network. Made in IBM Labs: IBM hot water-cooled supercomputer goes live at ETH Zurich. 02 July 2010. Retrieved from: 25/06/2014 World Wide Web. Available from: https://www-03.ibm.com/press/us/en/pressrelease/32049.wss

[18] US Department of Energy, Wireless Sensors Improve Data Centre Energy Efficiency. Federal Energy Management Program. September 2010; p. 1

[19] Sintes T. Events and listeners from Java World. 4 August 2000. [online]. Available from: http://www.javaworld.com/article/2077351/java-se/events-and-listeners.html

[20] Chaudhry MT, Ling TC, Hussain SA, Manzoor A. Minimizing thermal stress for data centre servers through thermal-aware relocation-research article. Scientific World Journal. 2014;**2014**, Article ID 684501. Received from: 2013, December, 02. [Accepted 2014, February, 12]. Published 2014, March, 31. Hindawi Publishing Corporation:1, 3-1, 4

[21] Sciacca C. (of IBM media relations). Zurich, IBM and ETH Zurich Unveil Plan to Build New Kind of Water-cooled Supercomputer. 20 June 2009 [online]. Available from: http://www.computerlekly.com/feature/IBM-builds-water-cooled-processor-for-Zurich-supercomputer

[22] Carbon Trust. Carbon Footprinting. April 2010. Retrieved 4/06/2014 World Wide Web. Available from: http://www.carbontrust.co.uk/cut-carbon-reducecosts/calculate/carbon-footprinting/pages/carbon-footprinting.aspx.

[23] Brunschwiler T, Smith B, Ruetsche E, Michel B. Toward zero-emission data centers through direct reuse of thermal energy. IBM Journal of Research and Development. Institute of Electrical and Electronics Engineers; 2009;**53**(3):1-13

Section 2

Energy Efficiency and Thermal Comfort

Chapter 5

Indoor Thermal Comfort from the Estimation Thermal Environment's Physical Variables in Temperate-Dry Bioclimate

Julio César Rincón-Martínez, Armando Núñez-de Anda and Francisco Fernández-Melchor

Abstract

Adverse thermal conditions alter the indoor habitability and, consequently, the occupants' health, performance, mood, and comfort. Although there are local regulations that provide thermal indicators for the indoor architectural design, these are usually unrelated to the climate, type of construction project, and psychophysiological adaptation of people. In this way, this chapter shows the thermal comfort ranges estimated from the thermal environment's physical variables, using the adaptive approach, for Ensenada City, Mexico (temperate-dry bioclimate). Surveys were applied to collect the subjective perception simultaneously with the measurement of black globe temperature, air temperature, relative humidity, and wind speed. Study sample was made up from students of a public university whose activity level is sedentary (1.2 met) and clothing is light (0.7 clo). The survey was designed based on the ISO 10551 and ANSI/ASHRAE 55 standards, while the physical measurement instruments were selected based on ISO 7726, managing to generate a class I database. The study was correlational and statistically analyzed with 3750 surveys from the average by thermal sensation intervals method. 16 comfort ranges were quantitatively and graphically estimated from the four physical variables analyzed in each of the four representative thermal periods of the city. These indicators offer objective knowledge for proper decision-making during the architectural design process and therefore for the thermal-energy efficiency of buildings.

Keywords: adaptive approach, adaptive thermal comfort, field study, indoor thermal comfort, measurement of environmental variables, psychophysiological adaptation, surveys

1. Introduction

The adverse conditions of the thermal environment can cause a direct affectation on the well-being, efficiency, and comfort of people. Studying this phenomenon from the adaptive approach allows obtaining quantitative estimates that respond to the thermal adaptation that people present to the local climate. The particular study of this phenomenon is caused by the need to have local thermal indicators that derive directly from the specific conditions of the climate, type of construction project, and psychophysiological adaptation of the inhabitants, since the architectural design is regularly based on local regulations that present a partial or total disconnection of the thermal parameters that promote the effective habitability of the spaces. However, there are national [1] and international [2–6] standards that establish the technical and methodological bases to estimate these thermal indicators based on the local particularities of each case. At the same time, different authors [7–34] have developed universal thermal comfort models [7–24] of graphic, mathematical, computer, or algorithmic type, which offer an approach to local thermal comfort [25–34] from outdoor weather and some endogenous factors of people (activity level and clothing).

So, the literature that documents this physical phenomenon [1–34] indicates that the mean radiant temperature, air temperature, relative humidity, and wind speed are the thermal environment's physical variables that present the greatest impact on the subjective perception that people manifest of the immediate environment, in addition to clothing and activity level.

However, the people's adaptation to the thermal environment also represents a contribution to the effective estimation of thermal comfort. Thermal adaptation is "the gradual decrease in the organism's response to repeated exposures to stimuli received from a specific environment" [35]. Therefore, the people's thermal perception depends on the physical and psychological sensations generated by the thermal environment, activity level, clothing, experience (thermal history), expectation, mean outdoor temperature, and time spent in the space [36].

Therefore, this chapter presents the adaptive thermal comfort estimated from the thermal environment's physical variables for the four representative thermal periods of Ensenada City, Mexico: cold period (December–March), thermal transition from the cold period to the warm period (April–June), warm period (July–September), and thermal transition from the warm period to the cold period (October–November). Additionally, it presents the adaptation actions taken by the people to get thermal comfort during these periods.

2. Methodology

The methodology is made up of the following sections: (1) Case study and target population, (2) Study periods, (3) Population sample, (4) Qualitative measurement instrument, (5) Environmental variables and physical measurement equipment, (6) Application of surveys, and (7) Data analysis.

2.1 Case study and target population

The study was carried out in Ensenada City, Mexico, geographically located at 31° 52′ latitude, −116° 40′ longitude, and 18 masl altitude [37] (**Figure 1**). The climate is extreme dry (BS0 ks(e)) [38], and bioclimate is temperate-dry [39]; the air temperature (AT) is 17.3°C, relative humidity (RH) is 75.8%, total rainfall (RF) is 217.3 mm, and the wind speed (WS) is 2.5 m/s, during a normalized year [40, 41].

The target population was made up of students from the Autonomous University of Baja California (UABC, acronym in Spanish), who present the following characteristics: age between 18 and 23 years, minimum time of residence in Ensenada City of 1 year, sedentary activity (1.2 met) [5], and moderate clothing (1.0 clo) [2].

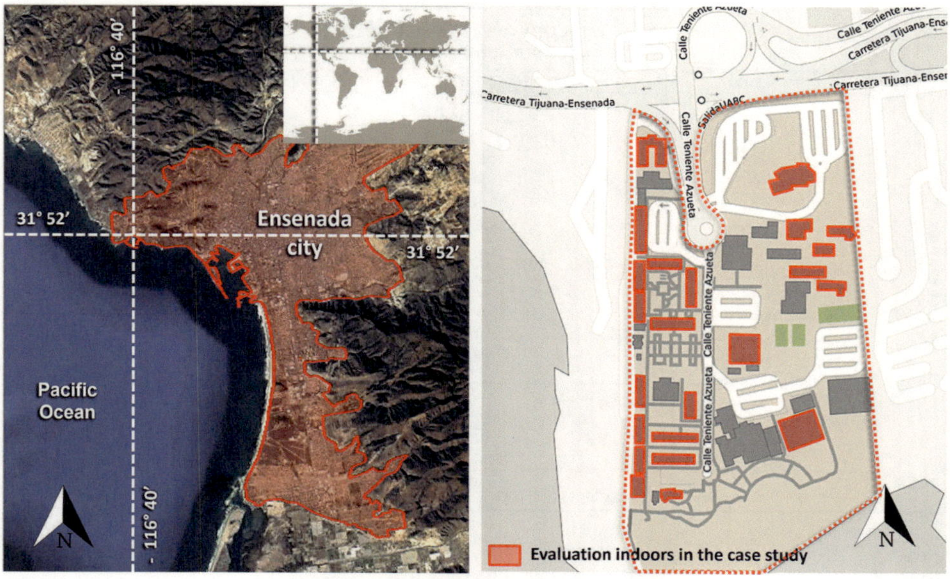

Figure 1.
Left. Geographic location and urban polygon of Ensenada City. Right. Study area and distribution of buildings used to carry out the study (Autonomous University of Baja California).

2.2 Study periods

The study was carried out in the four representative thermal periods of the city's climate: cold period (December to March), thermal transition from the cold period to the warm period (April to June), warm period (July to September), and thermal transition from the warm period to the cold period (October and November) (**Figure 2**). Hereinafter, the thermal transition periods will be referred to as thermal transition cold-warm and thermal transition warm-cold, respectively.

In this sense, the average climatic conditions that characterize each study period are presented in **Table 1** [40, 41].

2.3 Population sample

The population sample was designed with 95% confidence level and 5% confidence interval. Thus, the sample was estimated at 383 subjects for each study period; however, during the field work, it was possible to collect a sample greater than that required (**Table 2**). In total, 4153 observations were collected with the four study periods, of which 3750 observations had the certainty required to carry out the data analysis (1887 women and 1863 men).

2.4 Qualitative measurement instrument

The questionnaire was designed according to the one that was used by some referents specialized in the topic [30, 42–44] and what the national [1] and international [2–5] standards suggest in this regard. The questions related to the environment's hygrothermal sensation were based on the seven-point subjective scale shown in the ANSI/ASHRAE 55 [2] and ISO 10551 [3] standards, which was adapted as indicated in **Table 3**. The questionnaire's full version can be consulted in Appendix I.

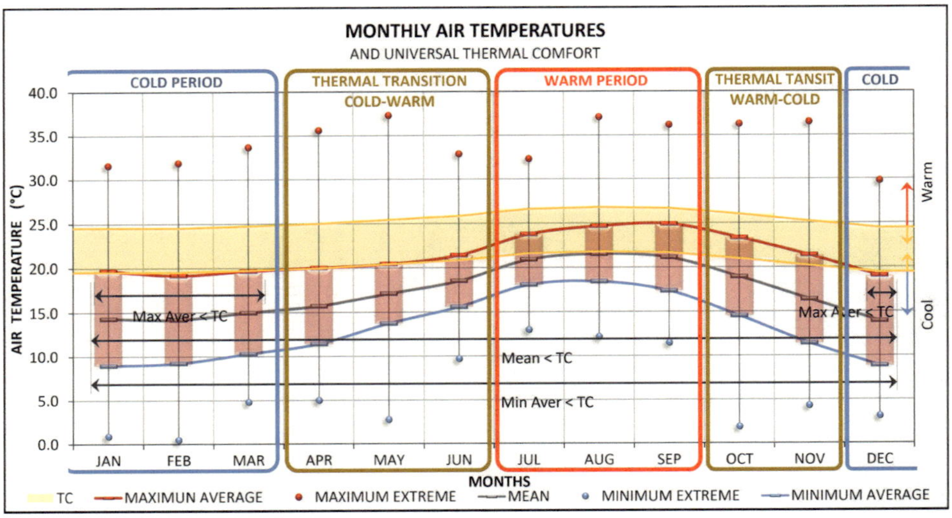

Figure 2.
Monthly thermal dynamics in a normalized year for Ensenada City.

Study period	Descriptors	AT (°C)	RH (%)	WS (m/s)	Rainfall (mm)
Cold	Average max	19.2	89.5	3.5	
	Mean	*14.2*	*70.7*	*2.8*	*49.1*
	Average min	9.2	46.2	2.0	
Thermal transition cold-warm	Average max	20.1	90.9	3.7	
	Mean	*17.1*	*78.1*	*2.7*	*7.1*
	Average min	13.7	63.0	2.0	
Warm	Average max	24.6	94.2	3.3	
	Mean	*21.5*	*86.1*	*2.0*	*1.5*
	Average min	18.4	69.3	1.3	
Thermal transition warm-cold	Average max	21.3	86.0	3.3	
	Mean	*16.4*	*67.0*	*2.6*	*13.3*
	Average min	11.4	44.8	1.8	

Table 1.
Climatic characterization for each study period [40, 41].

Study period	Sample collected	Sample analyzed	Women	Men
Cold	983	915	458	457
Thermal transition cold-warm	987	870	476	394
Warm	1365	1214	572	642
Thermal transition warm-cold	818	751	381	370

Table 2.
Stratification of the population sample for each study period and gender.

Thermal sensation	Hygric sensation	ANSI/ASHRAE 55 [2] ISO 10551 [3]	Adapted scale
Hot	Highly dry	+3	7
Warm	Dry	+2	6
Slightly warm	Slightly dry	+1	5
Neutral	Neutral	0	4
Slightly cool	Slightly wet	−1	3
Cool	Wet	−2	2
Cold	Highly wet	−3	1

Table 3.
Hygrothermal sensation scale used in the questionnaire/surveys.

2.5 Environmental variables and physical measurement equipment

Thermal environment's physical variables measured during the survey were: black-globe temperature (BGT), AT, RH, and WS as well as the clothing and activity level of the subjects. The measurement equipment used was a thermal environment monitor with three sensor arrays (3 M, model QUESTemp 36-3) (**Figure 3**). This equipment has a resolution of 0.1°C (BGT/AT), 0.1% (RH), and 0.1 m/s (WS); also, it has an accuracy of ±0.5°C (BGT/AT), ± 3.0% (RH), and ± 0.1 m/s (WS) [45]. The selection, distribution, and use of the instruments were carried out based on the ISO 7726 [6] and ANSI/ASHRAE 55 [2] standards, which allowed the creation of a class I database [46].

The evaluation indoor spaces were classrooms, computer rooms, and drawing workshops (**Figure 4**). When people were seated, the measurement sensors were set at 0.1, 0.6, and 1.1 m from the floor [2, 6]; but when people were semi-seated,

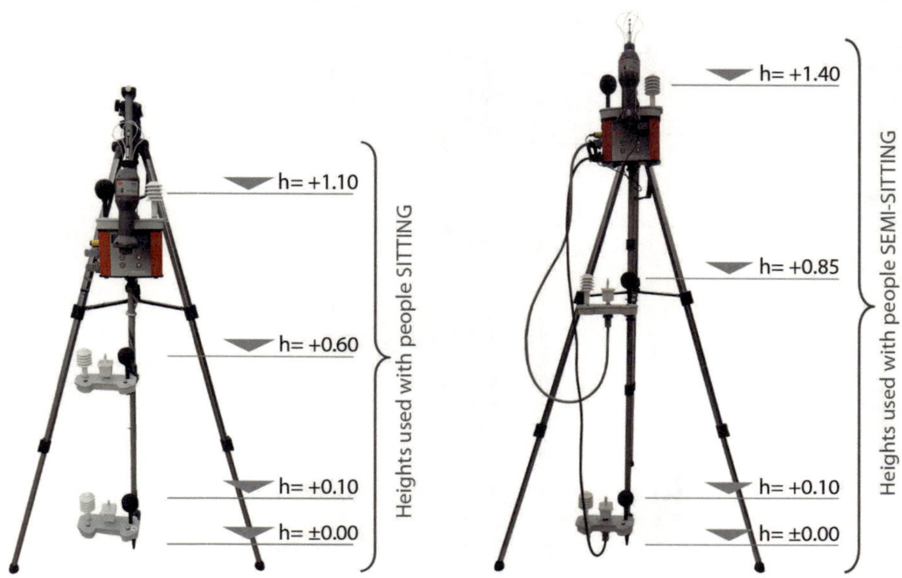

Figure 3.
Measuring equipment and heights used to measure the physical variables during the surveys.

the sensor heights were adapted to 0.1, 0.85, and 1.4 m, since it was not possible to consider them as sitting or standing (**Figure 3**).

2.6 Application of surveys

The surveys were applied in three-story buildings, built with concrete blocks and naturally ventilated (in isolated cases, natural ventilation was supplemented with air circulator fans). With this, the systematic procedure followed in the application of surveys is represented in **Figure 5**.

Technical stabilization of physical measurement equipment: 20 min before each evaluation, the physical measurement equipment was turned on in order to achieve stabilization of the recorded measurements.

Random selection of subjects to be evaluated: Two groups of students (of both genders) were randomly chosen daily during the extreme thermal times of the day: first, around 07:00 a.m. and second, around 03:00 p.m.

Distribution of physical measurement equipment: To begin the evaluation, the physical measurement equipment was distributed in the evaluation indoor space (location and heights), and the surveys were delivered to the study subjects (**Figure 3**).

Application of surveys and recorded variables: A survey leader conducted the surveys so that study subjects simultaneously responded to each question (**Figure 4**); at the same time, the technical support recorded the measurements of the thermal environment's physical variables. In this stage, those actions that the subjects carried out individually or collectively to adapt to the immediate environment were recorded. The duration of each survey was 18 min.

Figure 4.
Application of surveys in the evaluation indoor spaces. Left. Classroom with people sitting (heights: 0.1, 0.6, and 1.1 m). Right. Drawing workshop with people semi-sitting (heights: 0.1, 0.85, and 1.4 m).

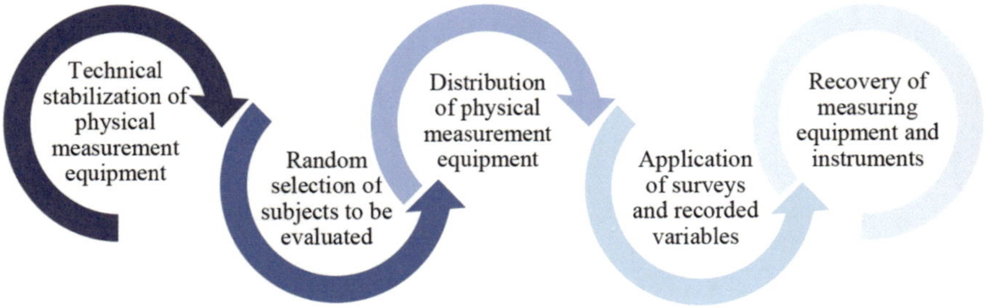

Figure 5.
Systematic procedure with which the field surveys were applied.

Recovery of measuring equipment and instruments: At the end of the survey, the measuring equipment and instruments were collected in an orderly manner in order to maintain a reliable parameter of data capture.

2.7 Data analysis

The collected data was analyzed by the averages for thermal sensation intervals (ATSI) method [47, 48]. This method groups the comfort votes by thermal sensation (TS) category to obtain the arithmetic mean of the physical variable registered in each of them and adds and subtracts ±2 standard deviation (SD) to each case in order to estimate the comfort range. Subsequently, it allows graphing the data pairs and plotting the simple linear regression: The neutral temperature (Nt) and comfort range (Cr) result from crossing them and the TS category number four.

Figure 6 exemplifies the statistical analysis by ATSI method to estimate thermal comfort by thermal environment's physical variable (neutral value and comfort range limits). Therefore, this data analysis had to be carried out 16 consecutive times: four

COLD PERIOD
Black Globe Temperature - Thermal Sensation
Processed evaluations: 915 (458 women, 457 men); activity level: 3 levels

SD	TS	Scale	-2 SD	-1 SD	Mean	+1 SD	+2 SD	Votes
1.9	Slightly warm	5	18.3	20.2	**22.1**	24.0	25.9	50
1.6	Neutral	4	17.1	18.7	**20.3**	21.9	23.5	442
1.5	Slightly cool	3	16.4	17.9	**19.4**	20.9	22.4	346
1.5	Cool	2	15.1	16.6	**18.1**	19.6	21.1	79

	-2 SD	-1 SD	Mean	+1 SD	+2 SD
Equation	*y = 0.96x -12.47*	*y = 0.85x -12.14*	*y = 0.76x -11.77*	*y = 0.69x -11.40*	*y = 0.63x -11.05*
r^2	*0.9849*	*0.9867*	*0.9814*	*0.9727*	*0.9625*
Neutral	**17.2**	**18.9**	**20.6**	**22.3**	**24.0**
Limits	*-3.4*	*-1.7*		*1.7*	*3.4*

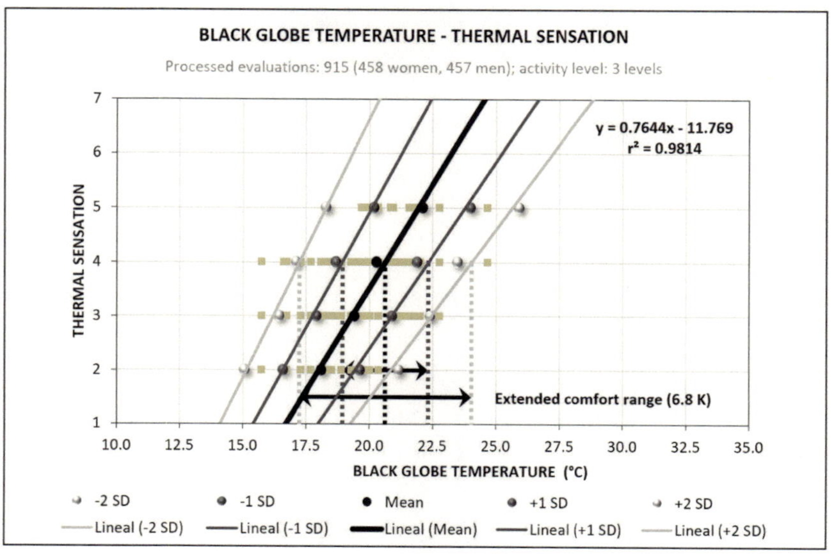

Figure 6.
Data analysis by ATSI method. This example only represents the analysis of BGT-TS, for the cold period.

physical variables (BGT, AT, RH, and WS) analyzed in each of the four study periods (cold, warm, and two thermal transition periods).

3. Results

The influence exerted by each thermal environment's physical variable on the TS was determined by the coefficient of determination (r^2). This allowed us to identify that BGT is the physical variable with the greatest influence on the TS, followed by AT, RH, and WS, respectively; thus, the results obtained are presented in this order. For this, **Table 4** shows the neutral value and comfort range of each thermal environment's physical variable to achieve and maintain the indoor thermal comfort in Ensenada City, México, throughout the year.

3.1 Thermal comfort from black globe temperature

According to **Figure 7**, thermal comfort from BGT is defined by a Nt of 20.6°C and a Cr of 17.2–24.0°C (thermal amplitude [Ta] of 6.8°C) for the cold period, a Nt of 21.5°C and a Cr of 18.6–24.5°C (Ta of 5.9°C) for the thermal transition cold-warm, a Nt of 24.5°C and a Cr of 22.4–26.6°C (Ta of 4.2°C) for the warm period, and a Nt of 23.0°C and a Cr of 20.1–25.9°C (Ta of 5.8°C) for the thermal transition warm-cold. In all cases, the upper and lower limits of the comfort range were equidistant from the Nt, which allows us to notice the same level of adaptation and tolerance at temperatures both above and below the Nt, establishing a thermal symmetry in each study period.

Additionally, the thermal dynamics that the comfort ranges present throughout the year suggests the psychophysiological adaptation by the subjects to achieve and maintain thermal comfort in view of the outdoor thermal conditions that occur in each study period (given the close relationship between naturally ventilated indoor and outdoor weather [36, 49, 50]). Therefore, due to outdoor thermal conditions of the

Study period	Values description	Neutral value and comfort range			
		BGT (°C)	AT (°C)	RH (%)	WS (m/s)
Cold	Upper limit	24.0	23.8	95.4	0.09
	Neutral	*20.6*	*20.3*	*60.8*	*0.02*
	Lower limit	17.2	16.8	27.5	0.00
Thermal transition cold-warm	Upper limit	24.5	22.7	83.7	0.22
	Neutral	*21.5*	*20.2*	*70.4*	*0.11*
	Lower limit	18.6	17.7	57.6	0.00
Warm	Upper limit	26.6	25.3	87.9	0.36
	Neutral	*24.5*	*23.1*	*74.1*	*0.17*
	Lower limit	22.4	20.9	60.2	0.00
Thermal transition warm-cold	Upper limit	25.9	25.7	80.0	0.11
	Neutral	*23.0*	*22.6*	*57.7*	*0.01*
	Lower limit	20.1	19.7	38.1	0.00

Table 4.
Thermal comfort from thermal environment's physical variable for each study period.

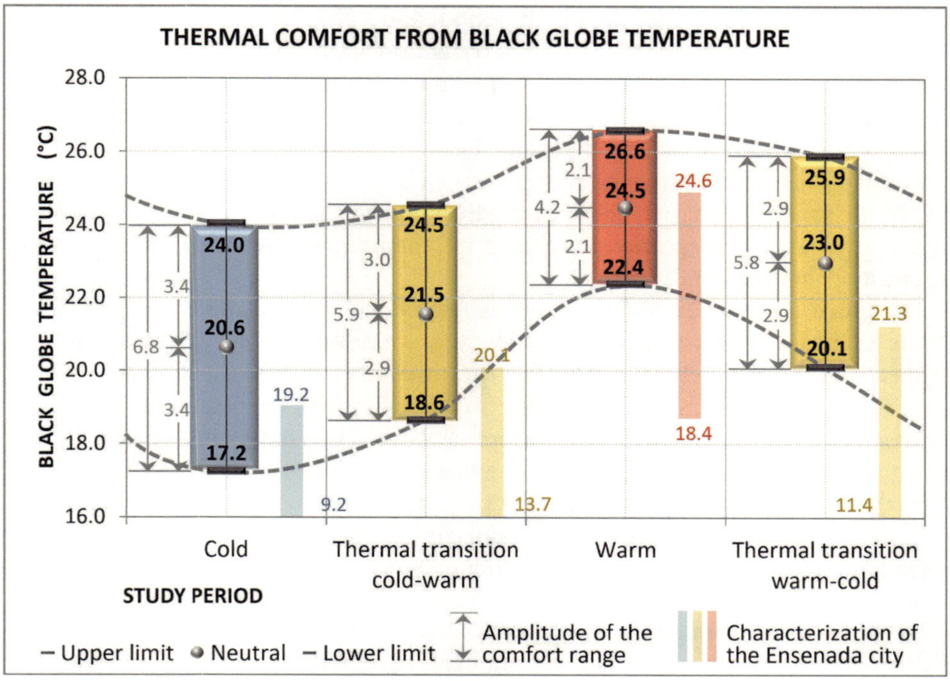

Figure 7.
Thermal comfort from BGT for each study period, contrasted with the thermal conditions of Ensenada City.

cold and warm periods, the comfort ranges for these periods resulted with the extreme values of the annual thermal dynamics, presenting the greatest thermal amplitude in the cold period (6.8°C) and the lowest thermal amplitude in the warm period (4.2°C); in contrast, the comfort ranges of the thermal transition periods presented thermal amplitudes equivalent to each other (5.9°C) and intermediate to the thermal amplitudes of the aforementioned periods. It is noteworthy in this last case that, although both periods are of thermal transition, each one presents specific magnitudes due to the direct and differentiated influence that the previous and subsequent thermal periods exert on each one of them (cold > warm; warm > cold), which would be equivalent to the thermal history and thermal expectation, respectively [36, 49, 50].

The above is attributed to the set of actions that people actively carry out to eventually achieve thermal comfort (opening and closing doors and windows, having drinks, using accessories or ventilation devices, etc.) as well as the level of clothing and the expectation generated in view of the approach of the next thermal period, achieving an efficient performance of their activities and favorable thermal conditions for their well-being.

Finally, the contextualization of these comfort ranges in the normalized conditions of the city (**Table 1**) allows us to identify the constant heating requirement throughout the year to achieve thermal comfort for people since only during the afternoons of each day, conditions that contribute to achieving thermal comfort occur naturally; however, during the rest of day (night, early morning, and morning), cold conditions occur (**Figure 7**).

3.2 Thermal comfort from air temperature

Thermal comfort from AT shown in **Figure 8** is defined by a Nt of 20.3°C and a Cr of 16.8–23.8°C (Ta of 7.0°C) for the cold period, a Nt of 20.2°C and a Cr of 17.7–22.7°C (Ta of 5.0°C) for the thermal transition cold-warm, a Nt of 23.1°C and a Cr of

20.9–25.3°C (Ta of 4.4°C) for the warm period, and a Nt of 22.6°C and a Cr of 19.7–25.7°C (Ta of 6.0°C) for the thermal transition warm-cold. As with the thermal comfort by BGT, in practically all cases, the upper and lower limits of the comfort range were equidistant from the Nt, which shows the same adaptation and tolerance at temperatures both above and below the Nt.

In this case, the thermal dynamics presented by the comfort ranges throughout the year is similar to that presented by the BGT comfort ranges, making visible the psychophysiological adaptation to achieve and maintain thermal comfort in view of the outdoor thermal conditions.

Therefore, again, the comfort ranges estimated for the cold and warm periods are the extremes of the annual thermal dynamics, presenting the highest thermal amplitude in the cold period (7.0°C) and the lowest thermal amplitude in the period warm (4.4°C); on the other hand, the thermal amplitude in the thermal transition periods remained intermediate to the aforementioned thermal amplitudes, evidencing a greater thermal amplitude in the thermal transition warm-cold (6.0°C) than in the thermal transition cold-warm (5.0°C). The analysis of this physical variable (AT) allows us to again identify the particularity of the thermal requirements for each thermal transition period, due to the influence exerted on them by the previous and subsequent thermal periods (cold>warm; warm>cold).

However, these comfort ranges contextualized in normalized climate of the city (**Table 1**) allow us to identify the constant heating requirement throughout the year to achieve thermal comfort since practically nights, early mornings, and mornings are cold (**Figure 8**), which could be solved with the architectural envelope of the buildings (shape, orientation, construction materials, thermal insulation, dimensions, heights, window-wall ratio, etc.) and voluntary/involuntary actions that contribute to the adaptation of the subjects to the immediate environment. It should be noted that the combination of the BGT/AT with the RH and the WS accentuates the thermal

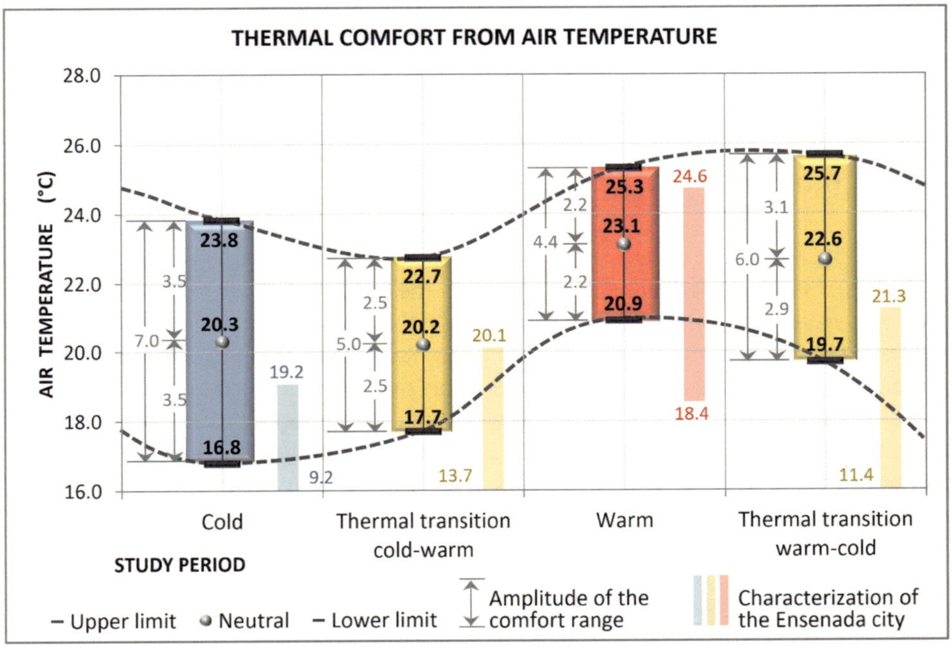

Figure 8.
Thermal comfort from AT for each study period, contrasted with the thermal conditions of Ensenada City.

sensation of the subjects throughout the year: in the warm period, the RH influences so that the weather is perceived as warmer and muggier, while in the cold period, the WS influences so that the weather is perceived as colder.

In summary, the annual thermal dynamics obtained with the thermal comfort from BGT is practically the same as that presented with the estimation of thermal comfort from AT, although with slightly higher values due to radiant temperature identified in the first case.

3.3 Thermal comfort from relative humidity

Thermal comfort from RH is defined by a neutral relative humidity (Nrh) of 60.8% and a Cr of 27.5–95.4% (hygric amplitude [Ha] of 67.9%) for the cold period, a Nrh of 70.4% and a Cr of 57.6–83.7% (Ha of 26.1%) for the thermal transition cold-warm, a Nrh of 74.1% and a Cr of 60.2–87.9% (Ha of 27.7%) for the warm period, and a Nrh of 57.7% and a Cr of 38.1–80.0% (Ha of 41.9%) for the thermal transition warm-cold (**Figure 9**). In none of the cases were the upper and lower limits of the comfort range equidistant from the Nrh, which means an asymmetric adaptation and tolerance by the subjects to humidity both above and below the Nrh. The Ha of the cold period is the largest of the annual hygric dynamics, in correspondence with the rainy season in the city, while the warm period presents the second-lowest Ha (but with values higher than 60.2%), due to the thermal damping required to minimize the thermal oscillation. It is noteworthy how the similarity with which the hygric values of the thermal transition cold-warm resulted with respect to those of the warm period.

Based on the hygric dynamics presented by the comfort ranges, it is possible to observe the psychophysiological adaptation that the subjects adopt to achieve and maintain indoor thermal comfort, which is linked to outdoor weather through natural ventilation [36, 49, 50]. In addition, from hygric amplitude with which each comfort range was estimated, the specific humidity requirements in each period can be determined to achieve thermal comfort, regardless of the outdoor hygric conditions. In other words, the hygric dynamics presented by the estimated comfort ranges are not linked to the hygric dynamics of Ensenada City but rather to the necessary requirements to achieve and maintain thermal comfort.

In this sense, **Figure 9** allows us to identify that the lower limit of the comfort ranges is less than the average minimum humidity that occurs outdoors in each period (**Table 1**), which indicates that the subjects would require a drier indoor environment to achieve thermal comfort during afternoons. Similarly, the upper limit of the comfort ranges in all periods, except in the cold period, is less than the average maximum humidity that occurs outdoors (**Table 1**), which again indicates the need for drier indoor environments during the mornings to achieve thermal comfort.

3.4 Thermal comfort from wind speed

Thermal comfort from WS is defined by a neutral wind speed (Nws) of 0.02 m/s and a Cr of 0.00–0.09 m/s (wind amplitude [Wa] of 0.09 m/s) for the cold period, a Nws of 0.11 m/s and a Cr of 0.00–0.22 m/s (Wa of 0.22 m/s) for the thermal transition cold-warm, a Nws of 0.17 m/s and a Cr of 0.00–0.36 m/s (Wa of 0.36 m/s) for the warm period, and a Nws of 0.01 m/s and a Cr of 0.00–0.11 m/s (Wa of 0.11 m/s) for the thermal transition warm-cold (**Figure 10**). Based on strictly statistical analysis, in all cases, the upper and lower limits of the comfort range were equidistant from the Nws; however, from the climatic point of view, the airflow does not present negative values in

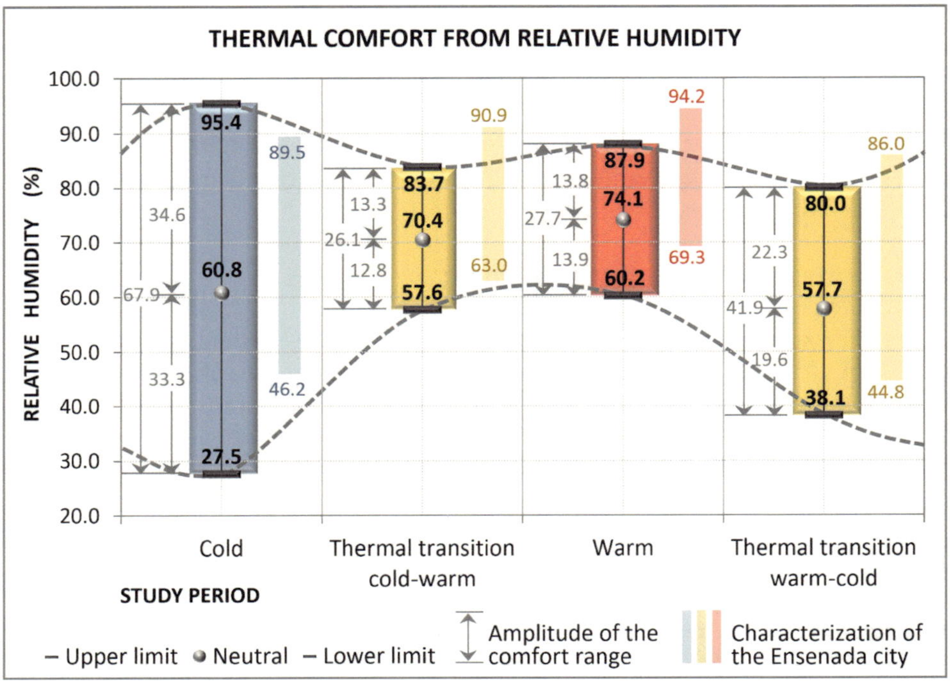

Figure 9.
Thermal comfort from RH for each study period, contrasted with the hygric conditions of Ensenada City.

the same direction, so they were replaced by 0.00 m/s. Thus, the limits of the comfort range were not equidistant from the Nws, which allows us to notice, from a phenomenological interpretation, a variable adaptation and tolerance at wind speeds both above and below the Nws, thus establishing a wind asymmetry in each study periods.

The wind dynamics presented by the comfort ranges derives from the wind requirements necessary to achieve thermal comfort, and not from the annual wind dynamics that occurs in Ensenada City. According to **Table 1**, the minimum WS in this case is 1.3 m/s, a speed above that required to achieve indoor thermal comfort. In this sense, and based on **Figure 10**, it is possible to appreciate that in the cold period, the lowest comfort range of the year (0.09 m/s) is required in order to preserve thermal comfort, avoid heat loss due to wind, and promote the cyclic air renewal; in contrast, the opposite occurs in the warm period; the highest comfort range (0.36 m/s) is required in order to achieve thermal comfort and promote heat loss during the afternoons. For its part, the comfort range estimated for thermal transitions resulted with an intermediate magnitude to those mentioned above, which, in addition, in each case, responds to the extreme thermal period to be transited (thermal expectation [36, 49, 50]); that is, the WS required for the thermal transition cold-warm (0.22 m/s) is greater than that required for the thermal transition warm-cold (0.11 m/s), corresponding to the approaching thermal period.

This requires strict control of natural ventilation in indoor spaces in order to reduce the risk of losing or gaining excessive heat and thus altering the thermal comfort achieved in each period.

The dynamics presented by the wind comfort ranges throughout the year also allows us to notice the psychophysiological adaptation that the subjects adopt to achieve thermal comfort in view of the direct influence of the outdoor weather (due to the close relationship between naturally ventilated indoor spaces and the outdoor weather [36, 49, 50]). The above supports the behavioral actions that subjects actively carry out to modify their

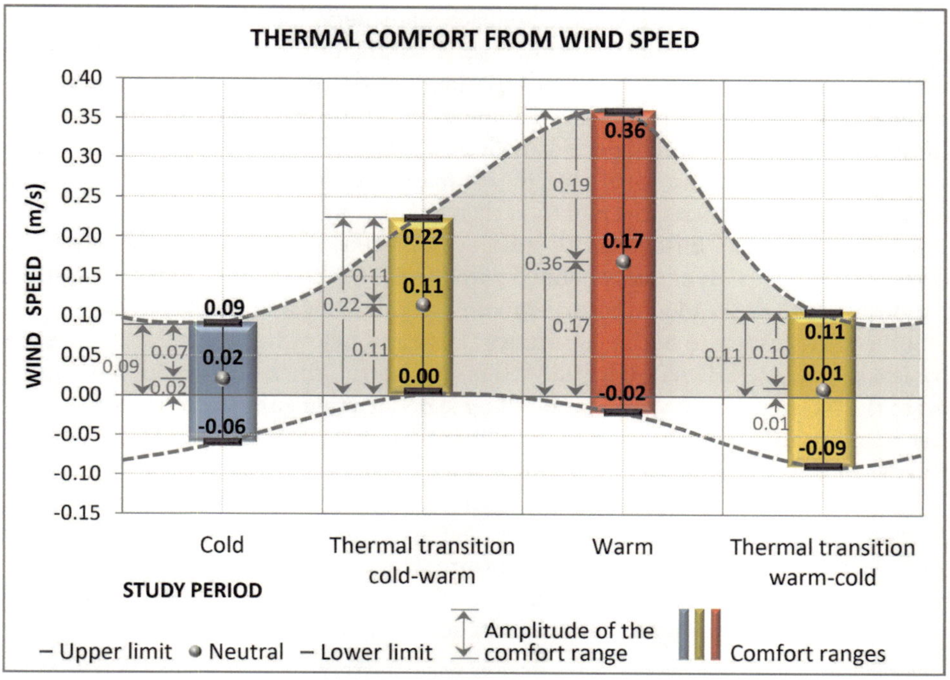

Figure 10.
Thermal comfort from WS for each study period.

immediate environment and clothing level to eventually achieve thermal comfort (opening/closing doors and windows, manual/mechanical ventilation, clothing, etc.), as well as the expectation that they generate in view of the approach of the next thermal period, getting an efficient performance and favorable thermal conditions for their well-being.

It is worth mentioning that with the wind analysis, the comfort ranges were not contrasted with the wind characterization for Ensenada city (**Figure 10**), since the minimum wind speed that occurs in this one is 1.3 m/s (**Table 1**), a speed well above that required in indoor spaces [16]. This suggests once again that wind control through windows, doors, and any opening is crucial to preserve indoor thermal comfort.

4. Conclusion

Thermal comfort is the eventual condition of people in which their psychophysiological perception achieves a balance with the immediate thermal environment, promoting a prolonged stay in the space, the efficient performance of their activities, and favorable conditions for their well-being in the short and medium terms. In spaces with a direct relationship to the outdoor, this phenomenon is mainly influenced by the environment's physical conditions; therefore, in order to preserve it, the active adaptation of subjects is decisive (e.g., conscious/unconscious attitudinal actions to partially and cyclically modify the immediate environment, their person, and their expectations).

In this sense, thermal comfort estimated from the adaptive approach for indoor spaces in Ensenada City, Mexico, is presented from the thermal environment's physical variables in **Table 4**; from BGT, in **Figure 7**; from AT, in **Figure 8**; from RH, in **Figure 9**; and, from WS, in **Figure 10**. The previous order represents the hierarchy with which the thermal environment's physical variables influenced the thermal sensation of subjects.

From the comfort ranges estimated for each study period, it was possible to appreciate the similarity of the annual thermal dynamics of the BGT with respect to the annual thermal dynamics of the AT. The slight variations presented are due to the factors that conceive each physical variable at each moment, which are less than 0.4°C in the thermal transition warm-cold and cold period and less than 1.8°C in the thermal transition cold-warm and warm period. In both cases, the estimated thermal comfort shows a constant heating requirement during the nights, early mornings, and mornings of the day, since the outdoor climatic of Ensenada city only cover this requirement during the afternoons; however, the thermal dynamics obtained with the BGT/AT comfort ranges have a strong relationship with the outdoor thermal dynamics by showing a similar behavior.

Subjects' tolerance to both magnitudes above and below the neutral value in each physical variable (BGT, AT, RH, and WS) is symmetric, since, in general, it is equidistant to the Nt, Nrh, and Nws. This allows us to deduce that subjects present the same level of acceptance to both magnitudes higher than and lower than the neutrality value, as long as the limits of the comfort range are not exceeded in each case.

Thermal transition periods present comfort ranges that are different from each other for each physical variable analyzed. This is due to the dependence that each one shows with respect to the thermal period from which they leave and to which they are approaching so that, although both periods are of thermal transition, each one presents comfort ranges with independent requirements.

Comfort ranges estimated with this study allow us to appreciate the adjustment that the subjects' thermal sensation adopts with the dynamics that the outdoor climate presents, motivating them to constantly search for psychophysiological adaptation that allows them to achieve local thermal comfort.

When the thermal environment presents continuous variability, the subjects carry out voluntary/involuntary actions that allow them to recover the thermal balance between the immediate environment and their organism. Common adaptation actions are: changing the clothing level, having drinks, changing position, mobility, shelter in a microclimate that promotes prompt acclimatization, use of natural conditioning devices (doors/windows), and eventual use of artificial conditioning mechanical devices.

Thus, thermal comfort estimated with this study derives from the group's thermal perception of the analyzed sample, which makes it possible to guarantee the adequate correspondence between the estimated thermal ranges and each of the study periods. These comfort ranges by physical variable represent local design parameters that, if considered during the conceptualization of the city's architectural projects, will directly impact the indoor habitability, the efficiency of the conditioning equipment, thermal comfort and the performance of the occupants, energy performance of buildings, passive architectural design, and sustainability, for example.

The results obtained with this study have a local scope and derive exclusively from the climate and population of Ensenada City, Mexico; however, they can be comparative references of other studies developed in similar physical conditions or a related area of knowledge.

Acknowledgements

Special thanks to Program for Teacher Professional Development (PRODEP, acronym in Spanish), Mexico, for funding the project "Indoor thermal comfort: A study in the temperate-dry bioclimate in Ensenada, Baja California", code UABC-PTC-607, within the call for support for the incorporation of *new full-time professors*, 2016, as well as the Autonomous University of Baja California (https://ror.org/05xwcq167), which registered the project with the code 402/395/E, for providing the facilities for the development of this research at its Ensenada campus.

Appendix I: Questionnaire used in the surveys

Front view

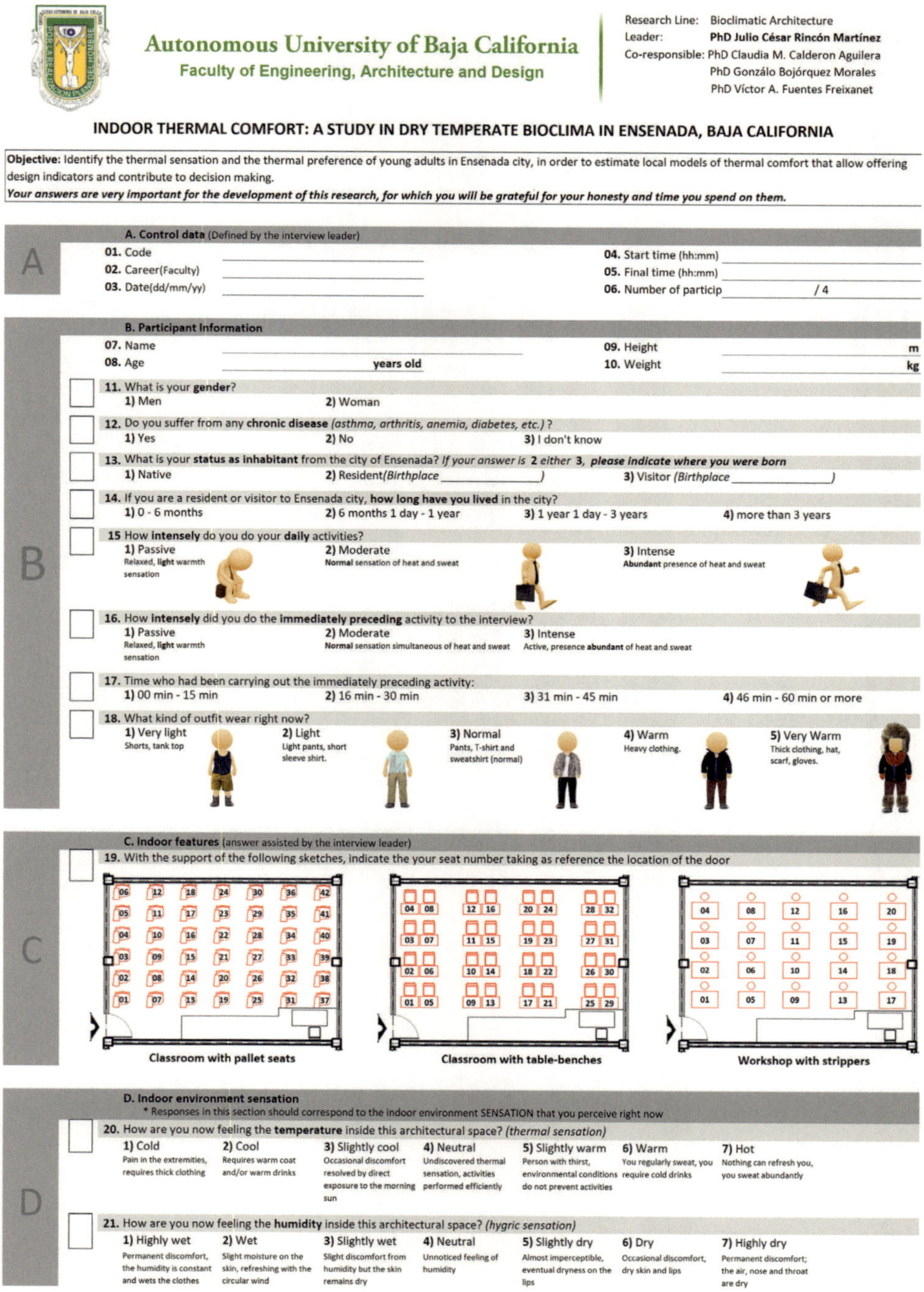

Autonomous University of Baja California
Faculty of Engineering, Architecture and Design

Research Line: Bioclimatic Architecture
Leader: **PhD Julio César Rincón Martínez**
Co-responsible: PhD Claudia M. Calderon Aguilera
PhD Gonzálo Bojórquez Morales
PhD Víctor A. Fuentes Freixanet

INDOOR THERMAL COMFORT: A STUDY IN DRY TEMPERATE BIOCLIMA IN ENSENADA, BAJA CALIFORNIA

Objective: Identify the thermal sensation and the thermal preference of young adults in Ensenada city, in order to estimate local models of thermal comfort that allow offering design indicators and contribute to decision making.
Your answers are very important for the development of this research, for which you will be grateful for your honesty and time you spend on them.

A. Control data (Defined by the interview leader)
01. Code
02. Career(Faculty)
03. Date(dd/mm/yy)
04. Start time (hh:mm)
05. Final time (hh:mm)
06. Number of particip / 4

B. Participant Information
07. Name
08. Age **years old**
09. Height **m**
10. Weight **kg**

11. What is your **gender**?
1) Men **2)** Woman

12. Do you suffer from any **chronic disease** *(asthma, arthritis, anemia, diabetes, etc.)*?
1) Yes **2)** No **3)** I don't know

13. What is your **status as inhabitant** from the city of Ensenada? *If your answer is* **2** *either* **3**, ***please indicate where you were born***
1) Native **2)** Resident*(Birthplace ______)* **3)** Visitor *(Birthplace ______)*

14. If you are a resident or visitor to Ensenada city, **how long have you lived** in the city?
1) 0 - 6 months **2)** 6 months 1 day - 1 year **3)** 1 year 1 day - 3 years **4)** more than 3 years

15 How **intensely** do you do your **daily** activities?
1) Passive — Relaxed, **light** warmth sensation
2) Moderate — **Normal** sensation of heat and sweat
3) Intense — **Abundant** presence of heat and sweat

16. How **intensely** did you do the **immediately preceding** activity to the interview?
1) Passive — Relaxed, **light** warmth sensation
2) Moderate — **Normal** sensation simultaneous of heat and sweat
3) Intense — Active, presence **abundant** of heat and sweat

17. Time who had been carrying out the immediately preceding activity:
1) 00 min - 15 min **2)** 16 min - 30 min **3)** 31 min - 45 min **4)** 46 min - 60 min or more

18. What kind of outfit wear right now?
1) Very light — Shorts, tank top
2) Light — Light pants, short sleeve shirt.
3) Normal — Pants, T-shirt and sweatshirt (normal)
4) Warm — Heavy clothing.
5) Very Warm — Thick clothing, hat, scarf, gloves.

C. Indoor features (answer assisted by the interview leader)
19. With the support of the following sketches, indicate the your seat number taking as reference the location of the door

Classroom with pallet seats **Classroom with table-benches** **Workshop with strippers**

D. Indoor environment sensation
* Responses in this section should correspond to the indoor environment SENSATION that you perceive right now

20. How are you now feeling the **temperature** inside this architectural space? *(thermal sensation)*
1) Cold — Pain in the extremities, requires thick clothing
2) Cool — Requires warm coat and/or warm drinks
3) Slightly cool — Occasional discomfort resolved by direct exposure to the morning sun
4) Neutral — Undiscovered thermal sensation, activities performed efficiently
5) Slightly warm — Person with thirst, environmental conditions do not prevent activities
6) Warm — You regularly sweat, you require cold drinks
7) Hot — Nothing can refresh you, you sweat abundantly

21. How are you now feeling the **humidity** inside this architectural space? *(hygric sensation)*
1) Highly wet — Permanent discomfort, the humidity is constant and wets the clothes
2) Wet — Slight moisture on the skin, refreshing with the circular wind
3) Slightly wet — Slight discomfort from humidity but the skin remains dry
4) Neutral — Unnoticed feeling of humidity
5) Slightly dry — Almost imperceptible, eventual dryness on the lips
6) Dry — Occasional discomfort, dry skin and lips
7) Highly dry — Permanent discomfort; the air, nose and throat are dry

Back view

D

22. How do you feel the **wind** right now? *(wind sensation)*

1) Windy	2) Slightly strong wind	3) Nice wind	4) Little wind	5) No wind
Strong wind, prevents sedentary activities	Flowing wind, causes some discomfort and moves light objects	Slight sensation of the wind on the skin, does not cause discomfort	There is no apparent movement of the air, slight suffocation is perceived	Heavy air, doesn't move, a lot of suffocation is perceived

23. What do you think of the **light** natural and/or artificial with which this space is illuminated? *(light sensation)*

1) Lousy	2) Bad	3) Regular	4) Good	5) Great
Extreme discomfort, activities can't be carried out normally	Permanent discomfort, requires adjustments to achieve adequate lighting levels	Slight discomfort but activities can be carried out without any change	Nice feeling regarding lighting levels, no adjustments required	The lighting is perfect, there is no vision problem or fatigue

24. The **noise** level from the immediate environment appears to you to be: *(hearing sensation)*

1) Very strong	2) Loud noise	3) Medium noise	4) Weak noise	5) No noise
Airport, railway, construction, etc.	Busy land route, concert, etc.	People living together, moderate vehicle traffic, etc.	Moderate chatter, barely audible music	Absolute silence

25. How do you perceive the **smell** of the indoor environment right now? *(olfactory sensation)*

1) Very unpleasant	2) Disgusting	3) Normal	4) Pleasant	5) Very nice
Unbearable and intense odors, you cannot stay in space	Certain unpleasantness of the smells but the activities can be carried out	Unnoticed sensation of some odor	Some acceptance of odors, contributes to activities	Total acceptance of odors, positively influences the space perception

26. How **tolerable** is the environment for you right now? *(personal tolerance)*

1) Perfectly tolerable	2) Tolerable	3) Neutral	4) Intolerable	5) Extremely intolerable
Excellent, totally pleasant and suitable for daily activities	The conditions are not totally to your liking, but they are sufficient for the activities	Indifferent, doesn't arouse any sensation, activities can be carried out	Requires adjustments to achieve the minimum conditions for the activities	Impossible to inhabit it, the conditions generate permanent displeasure

27. How would **consider** the indoor environment in general? *(personal acceptance)*

1) Generally acceptable 2) Generally unacceptable

E

E. Indoor environment preference

* It is very important that your answers derive from how ***YOU WOULD LIKE*** either ***YOU WOULD PREFER*** environmental conditions right now

28. How would you like to be right now about the **temperature**? *(thermal preference)*

1) Much more cold 2) Colder 3) Slightly cooler 4) Same as now 5) Slightly warmer 6) Warmer 7) Much warmer

29. If you could choose the indoor **humidity** level, would you decide that it was: *(hygric preference)*

1)Much wetter 2) Wetter 3) Slightly wetter 4) Same as now 5) Slightly drier 6) Drier 7)Much drier

30. Based on the indoor **wind** level right now, would you prefer it: *(wind preference)*

1) More wind 2) Same as now 3) Less wind

31. How would you prefer to see the **illumination** of space right now? *(light preference)*

1) More light 2) Same as now 3) Less light

32. If you could change the **sound** conditions that affect space, would you prefer them: *(sound preference)*

1) More noise 2) Same as now 3) Less noise

33. How would you like the smell you perceive right now to be presented? *(olfactory preference)*

1) More intense 2) Same as now 3) Less intense

F

F. Complementary information

34. What is your mood in relation to the environmental? *(Affective evaluation)*

1) Very bad	2) Bad	3) Slightly bad	4) Normal	5) Slightly good	6) Good	7) Very good
Angry, restless, depressed	Stressed, impatient	Bored, sleepy	Neither good nor bad: indifferent	Relaxed, concentrated	Happy, surprised	Excited, curious

35. In which thermal period of the year you present higher school performance? You can choose more than one option *(School performance)*

1) Cold period	2) Thermal transition cold-warm	3) Warm period	4) Thermal transition warm-cool
Winter	Spring	Summer	Autumn

36. In which thermal period of the year you presen higher performance labor (either domestic, if not working)? *(General/typical performance)*

1) Cold period	2) Thermal transition cold-warm	3) Warm period	4) Thermal transition warm-cool
Winter	Spring	Summer	Autumn

37. Notes:

Relevant or atypical situations that influence the environmental perception of the participant not considered in the questionnaire: Mild illness, moment of depression, physiological monthly period, exposure to certain pressure, worry or any mood other than ordinary, etc.

Author details

Julio César Rincón-Martínez[1*], Armando Núñez-de Anda[2]
and Francisco Fernández-Melchor[1]

1 Autonomous University of Baja California, Ensenada, Baja California, México

2 Autonomous University of Sinaloa, Culiacán, Sinaloa, México

*Address all correspondence to: julio.rincon@uabc.edu.mx

References

[1] Secretaría del Trabajo y Previsión Social [NOM-015-STPS-2001]. Condiciones Térmicas Elevadas o Abatidas-Condiciones de Seguridad e Higiene. México; 2001

[2] American Society of Heating. Refrigerating and Air Conditioning Engineers [ANSI/ASHRAE 55]. Thermal Environmental Conditions for Human Occupancy. Atlanta; 2017

[3] International Organization for Standardization [ISO 10551]. Ergonomics of Thermal Environment-Assessment of the Influence of the Thermal Environment Using Subjective Judgement Scales. Ginebra; 1995

[4] International Organization for Standardization [ISO 8996]. Ergonomics of the Thermal Environment-Determination of Metabolic Rate. 2nd ed. Ginebra; 2004

[5] International Organization for Standardization [ISO 7730]. Ergonomics of the Thermal Environment-Analytical Determination and Interpretation of Thermal Comfort Using Calculation of the PMV and PPD Indices and Local Thermal Comfort Criteria. 3rd ed. Ginebra; 2005

[6] International Organization for Standardization [ISO 7726]. Ergonomics of the Thermal Environment-Instruments for Measuring Physical Quantities. 2nd ed. Ginebra; 1998

[7] Houghten F, Yagloglou C. Determination of the Comfort Zone. American Society of Heating and Ventilating Engineers; 1923. p. 29

[8] Olgyay V. Arquitectura y Clima. Manual de Diseño Bioclimático para Arquitectos y Urbanistas. Barcelona: Gustavo Gili; 1963

[9] Fanger O. Thermal Comfort. 1st ed. New York: McGraw-Hill; 1972

[10] Auliciems A. Towards a psycho-physiological model of thermal perception. International Journal of Biometeorology. 1981;**25**:109-122

[11] Szokolay S. Climate analysis based on the psychometric chart. International Journal of Ambient Energy. 1986;**7**(4)

[12] de Dear R, Brager G, Cooper D. Developing an Adaptive Model of Thermal Comfort and Preference (final report on RP-884). USA: ASHRAE and Macquarie Research Ltd.; 1997

[13] Givoni B. Climate Considerations in Building and Urban Design. International Thomson Publishing, Inc; 1998

[14] Humphreys M, Nicol F. Understanding the adaptive approach to thermal comfort. ASHRAE Transactions, Technical Bulletin. 1998;**104**(1):991-1004

[15] Szokolay S. Introduction to Architectural Science: The Basis of Sustainable Design. London: Elsevier; 2004

[16] Evans J. The Comfort Triangles: A New Tool for Bioclimatic Design. Delft: Technische Universiteit Delft; 2007

[17] Humphreys M, Nicol F, Roaf S. Adaptive Thermal Comfort: Foundations and Analysis. 1st ed. London: Publisher Earthscan; 2015. DOI: 10.4324/9781315765815

[18] Földváry-Ličinaa V et al. Development of the ASHRAE global

thermal comfort database II. Building and Environment. 2018;**142**:502-512. DOI: 10.1016/j.buildenv.2018.06.022

[19] Montazami A, Gaterell M, Nicol F, Lumley M, Thoua C. Developing an algorithm to illustrate the likelihood of the dissatisfaction rate with relation to the indoor temperature in naturally ventilated classrooms. Building and Environment. 2017;**111**:61-71. DOI: 10.1016/j.buildenv.2016.10.009

[20] Cheung T, Schiavon S, Parkison T, Li P, Brager G. Analysis of the accuracy on PMV-PPD model using the ASHRAE global thermal comfort database II. Building and Environment. 2019;**153**:205-2017. DOI: 10.1016/j.buildenv.2019.01.055

[21] Liu S, Schiavon S, Prasanna-Das H, Jin M, Spanos C. Personal thermal comfort models with wearable sensors. Building and Environment. 2019;**162**:1-17. DOI: 10.1016/j.buildenv.2019.106281

[22] Loomans MGLC, Mishra A, Kooi L. Long-term monitoring for indoor climate assessment-the association between objective and subjective data. Building and Environment. 2020;**179**:1-12. DOI: 10.1016/j.buildenv.2020.106978

[23] Luo M, Xie J, Yan Y, Ke Z, Yu P, Wang Z, et al. Comparing machine learning algorithms in predicting thermal sensation using ASHRAE comfort database II. Energy and Building. 2020;**210**:1-16. DOI: 10.1016/j.enbuild.2020.109776

[24] Ji W, Zhu Y, Cao B. Development of the predicted thermal sensation (PTS) model using the ASHRAE global thermal comfort database. Energy and Building. 2020;**211**:1-12. DOI: 10.1016/j.enbuild.2020.109780

[25] Nicol F, Jamy G, Sykes O, Humphreys M, Roaf I, Hancock M. A Survey of Thermal Comfort in Pakistan toward New Indoor Temperature Standards. United King: Oxford Brookes University; 1993

[26] Boerstra A, Kurvers S, Van der Linden A. Thermal comfort in real live buildings: Proposal for a New Dutch Guideline. In: Levin H, editor. Proceedings of the 9th International Conference on Indoor Air. 1st ed. 2002. pp. 629-6342

[27] Ruíz R. Estándar Local de Confort Térmico para la Ciudad de COLIMA. México: Universidad de Colima; 2007

[28] Mishra A, Ramgopal M. A thermal comfort field study of naturally ventilated classrooms in Kharagpur, India. Building and Environment. 2015;**92**:396-406. DOI: 10.1016/j.buildenv.2015.05.024

[29] Jindal A. Thermal comfort study in naturally ventilated school classrooms in composite climate of India. Building and Environment. 2018;**142**:34-46. DOI: 10.1016/j.buildenv.2018.05.051

[30] Rincón-Martínez, J. Confort Térmico en Interiores: Estimación con los Enfoques Adaptativo y Predictivo. México: Universidad Autónoma de Baja California; 2019. Available from: https://libreriauabc.com/products/confort-termico-en-interiores-estimacion-con-los-enfoques-adaptativo-y-predictivo-fragmento-de-julio-cesar-rincon-martinez-confort-termico-en-bioclima-semi-frio-estimacion-a-partir-de-los-enfoques-de-estudio-adaptativo-y-predictivo [Accessed: December 17, 2022]

[31] Buonocorea C, De Vecchib R, Scalcoc V, Lamberts R. Thermal preference and comfort assessment in air-conditioned and naturally-ventilated university classrooms under hot and

humid conditions in Brazil. Energy and Building. 2020;**211**:1-13. DOI: 10.1016/j.enbuild.2020.109783

[32] Rincón-Martínez J, Martínez-Torrez K, González-Trevizo M, Fernández-Melchor F. Modelos Matemáticos para Estimar el Confort Térmico Adaptativo en Espacios Interiores: Un Estudio en la Transición Térmica de Ensenada, B.C. Ingeniería. 2020;**24**(1):1-17. Available from: https://www.revista.ingenieria.uady.mx/ojs/index.php/ingenieria/article/view/186 [Accessed: November 21, 2022]

[33] Alibaba HZ. Adaptive thermal comfort of an Office for Energy Consumption-Famagusta Case. In: Bienvenido-Huertas D, editor. Nearly Zero Energy Building (NZEB) - Materials, Design and New Approaches. London: IntechOpen; 2021. DOI: 10.5772/intechopen.101077

[34] Samsuddin R, R., Baharuddin, H. & Rosady, M. Proposed model of neutral temperature equation for adaptive thermal comfort in student flats units in the tropics. Civil Engineering and Architecture. 2021;**9**(2):477-492. DOI: 10.13189/cea.2021.090220

[35] Nikolopoulou M, Steemers K. Thermal comfort and psychological adaptation as a guide for designing urban spaces. Energy and Buildings. 2003;**35**:95-101. DOI: 10.1016/S0378-7788(02)00084-1

[36] Humphreys M, Nicol F. The validity of ISO-PMV for predicting comfort votes in every-day thermal environments. Energy and Buildings. 2002;**34**:667-684. DOI: 10.1016/S0378-7788(02)00018-X

[37] Instituto Nacional de Estadística y Geografía. Prontuario de Información Geográfica Municipal de los Estados Unidos Mexicanos: Ensenada, Baja California (Clave geoestadística 02001). México: INEGI; 2009

[38] García E. Modificaciones al Sistema de Clasificación Climática de Köppen [para Adaptarlo a las Condiciones de la República Mexicana]. México: Instituto de Geografía, Universidad Nacional Autónoma de México; 2004

[39] Fuentes V, Figueroa A. Criterios de Adecuación Bioclimática en la Arquitectura. México: Instituto Mexicano del Seguro Social; 1990

[40] Servicio Meteorológico Nacional. Datos Climáticos Registrados por la Estación Meteorológica Automática (BC-02 Ensenada), periodo 2000-2017. México: SMN-CONAGUA; 2017

[41] Huerta-Tapia Y. Manual de diseño pasivo para el arquitecto: Un Reflejo del Análisis Climático y Bioclimático de Ensenada, Baja California. México: Universidad Autónoma de Baja California; 2018

[42] González F. Zona de Confort Higrotérmico para Adultos Jóvenes de la Ciudad de México [thesis]. México: Instituto Politécnico Nacional; 2012

[43] Gómez-Azpeitia G, Ruiz R, Bojórquez G, Romero R. Confort Térmico y Ahorro de Energía en la Vivienda Económica en México, Regiones de Clima Cálido Seco y Húmedo. México: Comisión Nacional del Fondo para la Vivienda; 2007

[44] Stoops J. The Physical Environment and Occupant Thermal Perceptions in Office Buildings: An Evaluation of Sampled Data from Five European Countries. Suecia: Chalmers University of Technology; 2001

[45] 3M. Monitor de Ambiente Térmico QT 36-3 [Internet]. 2023. Available

from: https://www.3m.com.mx/3M/es_MX/p/d/v000183921/ [Accessed: January 01, 2023].

[46] Brager G, de Dear R. Thermal adaptation in the built environment: A literature review. Energy and Buildings. 1998;**27**(1):83-96. DOI: 10.1016/S0378-7788(97)00053-4

[47] Rincón-Martínez J. Basic methods used for data analysis in adaptive thermal comfort studies. Ingeniería, Investigación y Tecnología. 2023;**01**:1-17. DOI: 10.22201/fi.25940732e.2023.24.1.002

[48] Gómez-Azpeitia G, Ruiz R, Bojórquez G, Romero R. Monitoreo de Condiciones de Confort Térmico: Reporte Técnico (Producto 3), CONAFOVI 2004-01-20. México: Comisión Nacional del Fondo para la Vivienda; 2007

[49] Auliciems A, de Dear R. Thermal adaptation and variable indoor climate control. In: Auliciems A, editor. Human Bioclimatology. Advances in Bioclimatology. 1st ed. Berlin: Springer; 1998. pp. 61-86. DOI: 10.1007/978-3-642-80419-9_3

[50] Auliciems A, Szokolay S. Thermal comfort. In: Auliciems A, Szokolay S, editors. Notes of Passive and Low Energy Architecture International. Brisbane: PLEA-University of Queensland; 1997. pp. 3-65 Available from: http://www.plea-arch.org/index.php/plea-publications/ [Accessed: January 01, 2023]

Chapter 6

Energy Efficiency, Thermal Comfort, and Quality of Natural Ventilation Strategies for Classrooms

Paula Scherer, Daiana de Oliveira Fauro and Giane de Campos Grigoletti

Abstract

Classrooms are long-term environments, in which thermal comfort is essential for a good teaching and learning process. This research presents 16 natural ventilation strategies for classrooms related to energy efficiency, thermal comfort, and the quality of natural ventilation, for regions with a humid subtropical climate, represented by the Brazilian city of Santa Maria. Computer simulations were carried out with the Ansys CFX and EnergyPlus software, in addition to thermal comfort criteria recommended by the adaptive model of ASHRAE 55/2017, where the hours spent in thermal comfort, degree-hours of discomfort, indoor air velocity, air renewal rates, and air humidity were analyzed. The results indicate the most favorable natural ventilation strategies for classrooms located in a developing country in a humid subtropical climate, showing that these can contribute to the improvement of the quality of natural ventilation compared to the conventional constructive solutions used with simple strategies.

Keywords: natural ventilation, thermal comfort, classrooms, energy efficiency, air quality

1. Introduction

Natural ventilation aims to exchange air between the internal and external environments without the use of mechanical systems and at the lowest energy consumption. Constant air exchange is an essential bioclimatic architecture parameter and aims to improve users' thermal comfort and health in indoor environments.

Natural ventilation is fundamental in classrooms, and its importance has increased due to the COVID-19 pandemic. Air renewals are essential for indoor air quality, not only to reduce CO_2 concentration but also to avoid the risk of transmission of viral diseases [1–4]. In addition, the quality of the indoor environment influences children's interest in learning. All these factors point to the importance of using strategies that provide thermal comfort and adequate air renewal rates [5, 6].

Passive strategies based on natural ventilation must be appropriate to the climate and user's habits as well as to the economic context. Considering Brazil, a country with a large territory and different climate conditions, passive natural ventilation strategies should be well chosen, both for their suitability to the climate and due to economic restrictions, since frequently public schools do not have air conditioning systems. For example, a large study in Brazilian countryside, which evaluated thermal comfort and natural ventilation in standardized schools located in different climates [7], demonstrated the importance of choosing the appropriate strategies according to the climate context, criticizing the use of standard solutions. Considering a tropical climate in high altitudes, the use of solar chimneys for ventilation in classrooms was adequate as demonstrated by computational simulations and in situ measurements, improving user's thermal comfort [8]. Regarding equatorial climate, a study demonstrated that window frame designs could improve natural ventilation and children's thermal comfort through simple solutions. Also, for equatorial climates [9], ventilated sills could increase air velocity and thermal performance in classrooms. All these mentioned studies indicate the difference between solutions according to different contexts in Brazil.

In relation to the evaluation of natural ventilation, in order to guarantee indoor air quality, variables such as suitable air renewal rates are considered. The World Health Organization (WHO) recommends, for non-residential buildings such as schools, 10 liters per second per person and a relative humidity between 60 and 80%. This air renewal rate is higher than the Brazilian standardized value that is still not adjusted to new demands from COVID-19 pandemic.

There are several ways to provide natural ventilation from the use of simple windows (one-ventilated or cross-ventilated) to devices that use the temperature difference of the air layers to force its movement, such as chimney effect.

Which strategy is the most recommended is based on other criteria besides the recommended air renewal rates, such as users' thermal comfort, especially for climates with periods characterized as cold, such as the humid subtropical, in which ventilation can cause great thermal discomfort due to low temperatures. Thermal comfort can be expressed by the number of hours in comfort with respect to the total number of occupied hours, besides the system energy efficiency, since when natural ventilation is not sufficient, artificial air conditioning should be turned on [10]. In this case, the degree-hour method for heating and cooling can be used to assess energy efficiency. In addition, air velocity at the height of users' body should be considered in order to avoid discomfort and undesirable effects such as flying of papers, important to consider in a classroom environment [11].

Taking in consideration the importance of natural ventilation in school environments in an era of pandemic and the need to find systems with better energy efficiency and thermal comfort according to climate and economic aspects, this chapter presents the evaluation of 16 passive strategies for natural ventilation for a subtropical climate in developing countries such as Brazil. The classroom evaluated is intended for children in the age range 6–14 years and located in a humid subtropical climate with four well-defined seasons, which is represented by a southern Brazilian city, named Santa Maria.

The evaluation was based on computational simulations with CFX and EnergyPlus, as well as air renewal rates and relative air humidity recommended by WHO, thermal comfort expressed by occupied hours in thermal comfort, and ventilation strategies energy efficiency expressed by degree-hours for heating and cooling, according to ASHRAE [10].

2. Background and method

2.1 Climate

Santa Maria, which represents the studied humid subtropical climate, is located 300 km from the seaside, at a latitude of 29.70 S, longitude of 53.82 W, and average altitude of 115 m. The file adopted is TMY2. The city is characterized by mean annual maximum and minimum temperatures of 25.4 and 14.9°C, respectively. The mean annual maximum and minimum relative air humidity is 99 and 32%, respectively. The air velocity presents mean maximum and minimum 7.33 and 0.05 m/s, respectively. The predominant wind directions are east and southeast [12]. **Table 1** presents Santa Maria climatological normals. **Figure 1** illustrates the wind frequency in Santa Maria.

Month	01	02	03	04	05	06	07	08	09	10	11	12	Year
Max. temp. (°C)	41.0	41.0	40.0	39.6	34.1	31.2	32.0	35.2	36.8	37.9	40	40.2	41.0
Mean max. temp. (°C)	31.0	30.2	29.1	26.0	22.0	19.8	19.2	21.8	22.5	25.3	28.0	30.4	25.4
Min. temp. (°C)	9.4	9.4	5.9	2.5	0.1	−2.6	−2.9	−2.0	−0.6	3.4	5.2	7.2	−2.9
Mean min. temp. (°C)	20.1	19.7	18.3	15.2	12.1	10.4	9.4	10.8	12.3	15.0	16.5	18.7	14.9
Rainfall (mm)	166	132	142	151	137	133	147	11	155	203	136	162	1.778
Relative humidity (%)	74	77	79	81	84	84	83	79	79	77	71	70	78
Sun hours	243	206	213	179	156	127	147	166	159	181	225	250	2.249

Table 1.
Santa Maria climatological normals, adapted from Ref. [12].

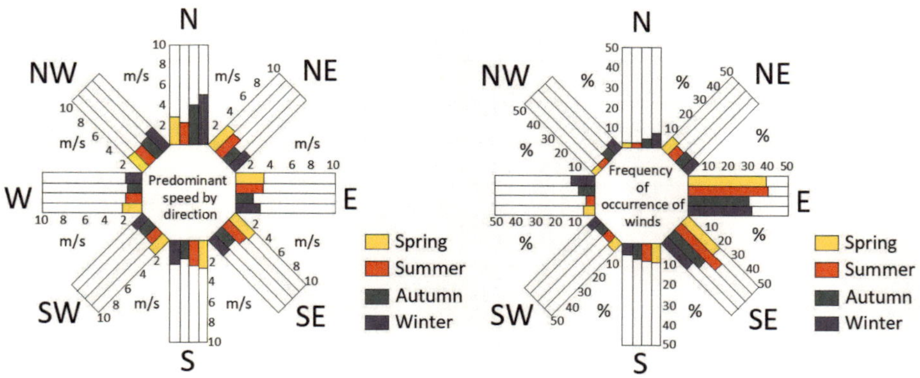

Figure 1.
Wind frequency in Santa Maria, translated from Ref. [13].

2.2 Classroom standard and proposed strategies

The classroom model adopted in the simulations is recommended by the Brazilian government for public schools and is widely used throughout the country, with minor adaptations according to the climate, mainly for walls. The classroom has an area equal to 45.00 m^2 and ceiling height 2.80 m. Its maximum capacity is 36 students and 1 teacher (37 users). **Figure 2** illustrates the classroom geometry. The roof was represented as transparent for the visualization of the indoor environment.

The construction technology is based on the concrete slab and ceramic floor, rectangular 6-hole hollow/porotherm clay block for walls (9 cm × 14 cm × 24 cm), single glass for windows, and plywood panel for door. The physical properties of the materials are presented in **Table 2**, considering thickness (e), thermal conductivity (λ), specific weight (ρ), specific heat (c), thermal emittance (ε), solar absorbance (α), transmissivity (α), and reflectivity.

The standard classroom has a single-sided ventilation, window size 10.96 m^2, and sill equal to 1.00 m from floor. Window orientation is east in order to capture the

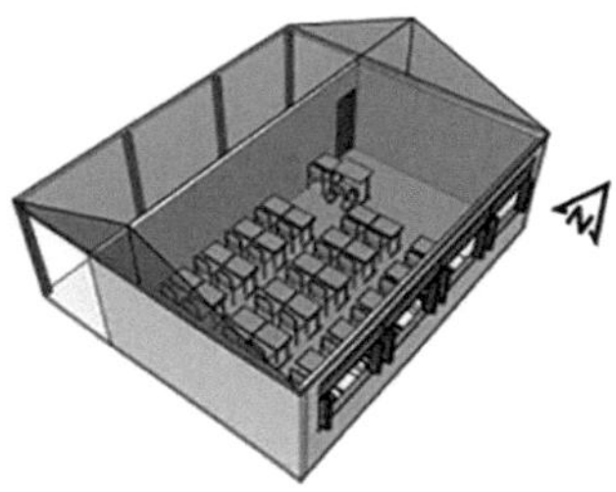

Figure 2.
Classroom geometry.

Component	e (cm)	λ (W/m°C)	ρ (kg/m³)	c (kJ/kgK)	ε (W/mK)	α
Ceramic floor	0.750	1.05	2000	0.92	0.90	0.20
Mortar	2.00	1.15	2000	1.00	0.90	0.70
Concrete	9.85	1.75	2200	1.00	0.90	0.80
Mortar (2x)	2.50	1.15	2000	1.00	0.90	0.20
Clay (2x)	1.34	0.90	1600	0.92	0.90	0.70
Air chamber	6.32	0.36	—	—	—	—
Clay	1.00	1.05	2000	0.92	0.90	0.20
Air chamber	25.00	1.19	—	—	—	—
Concrete	4.00	1.75	2200	1.00	0.90	0.20
EPS + concrete	7.00	0.223	2000	1.00	0.90	0.20
Mortar	1.00	1.15	2000	1.00	0.90	0.20
Plywood	3.50	0.15	550	2.30	0.90	0.70
Single glass	0.30	—	0.09	—	—	0.08

Table 2.
Components' physical properties [14–16].

prevailing regional winds. The window is divided into two parts: the bottom is an awning panel, and the top is a maxim-ar panel (with 90° aperture).

Models	
1. Awning single-sided ventilation, east orientation, area 10.96 m^2	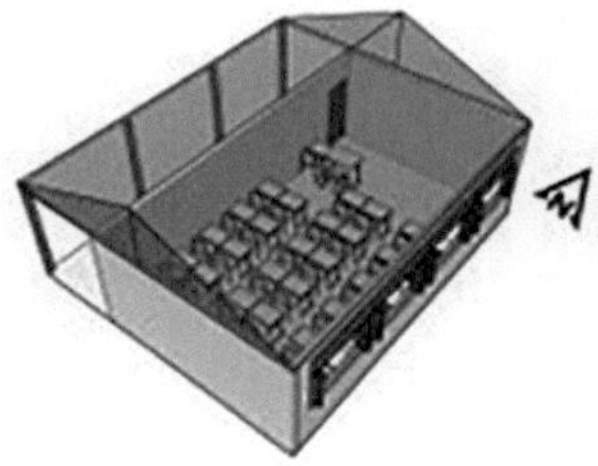
2. Awning single-sided ventilation, east orientation, area 9.28 m^2 + east-facing maxim-ar upper opening for permanent ventilation, area 1.68 m^2	
3. Cross-ventilation (east–west), area 13.82 m^2	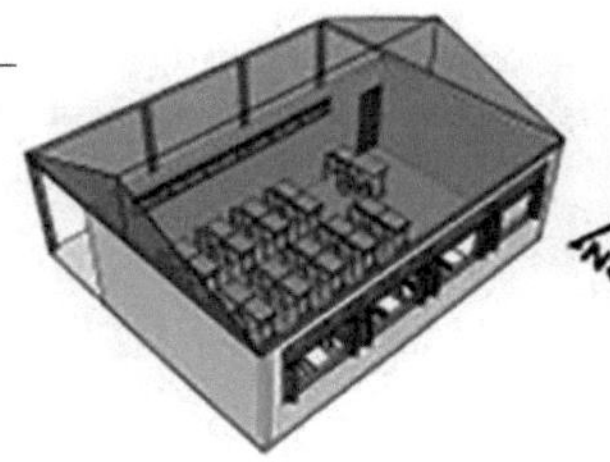
4. Cross-ventilation (east–west), area 9.28 m^2 + east-facing maxim-ar upper opening for permanent ventilation, area 1.68 m^2	
5. Cross-ventilation (east–west) + stack effect using openings on the roof of the west-facing corridor, area 13.82 m^2	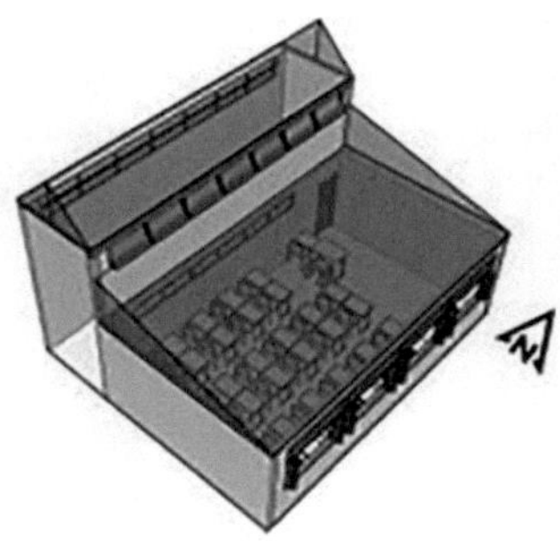
6. Cross-ventilation (east–west) + stack effect using openings on the roof of the west-facing corridor (area 12.14 m^2) + east-facing maxim-ar upper opening for permanent ventilation, area 1.68 m^2	
7. Awning single-sided ventilation, east orientation + two west-facing solar chimneys, area 11.68 m^2	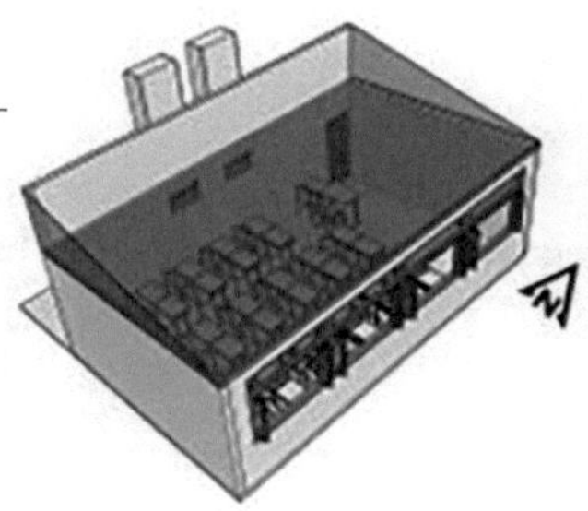
8. Awning single-sided ventilation, east orientation + two west-facing solar chimneys, area 10.00 m^2 + east-facing maxim-ar upper opening for permanent ventilation, area 1.68 m^2	
9. Awning single-sided ventilation, north orientation, area 10.96 m^2	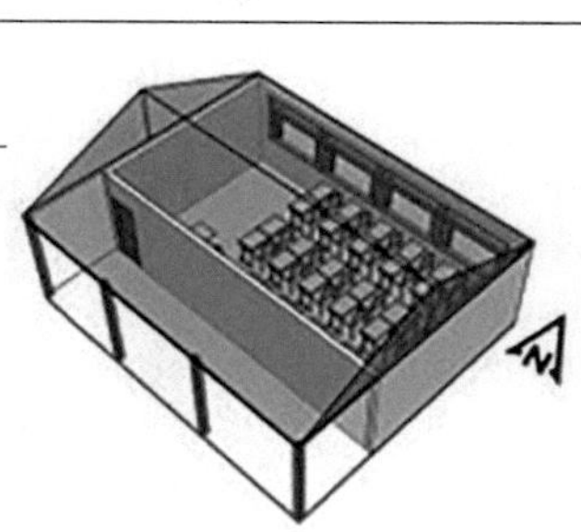
10. Awning single-sided ventilation, north orientation, area 9.28 m^2 + north-facing maxim-ar upper opening for permanent ventilation, area 1.68 m^2	

Models	
11. Cross-ventilation (north–south), area 13.82 m^2	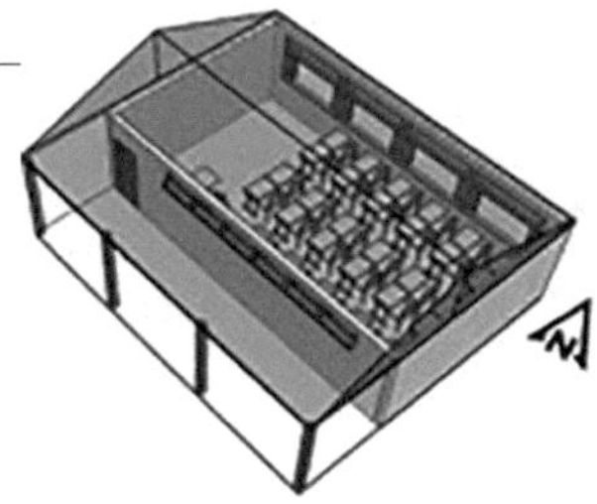
12. Cross-ventilation (north–south) 9.28 m^2 + south-facing maxim-ar upper opening for permanent ventilation, area 1.68 m^2	
13. Cross-ventilation (north–south) + stack effect using openings on the roof of the south-facing corridor, area 13.82 m^2	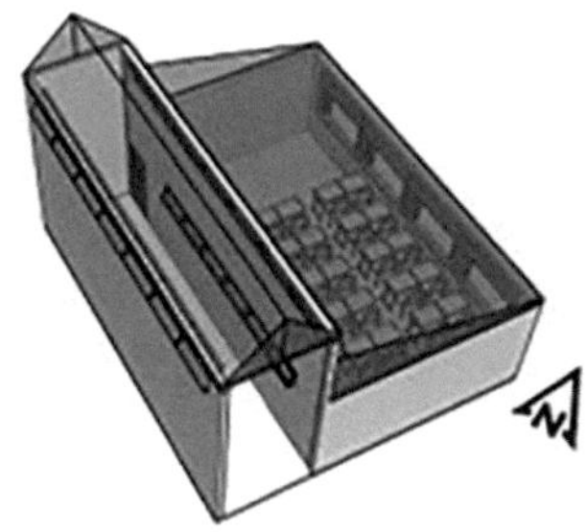
14. Cross-ventilation (north–south) + stack effect using openings on the roof of the south-facing corridor (area 12.14 m^2) + north-facing maxim-ar upper opening for permanent ventilation, area 1.68 m^2	
15. Awning single-sided ventilation, north orientation + two south-facing solar chimneys, area 11.68 m^2	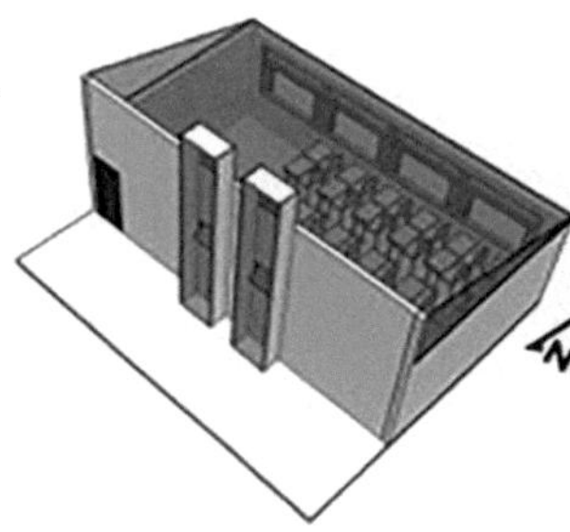
16. Awning single-sided ventilation, north orientation + two south-facing solar chimneys, area 10.00 m^2 + north-facing maxim-ar upper opening for permanent ventilation, area 1.68 m^2	

Table 3.
Simulated strategies.

In addition to the standard solution described above, 15 window models were proposed as illustrated in **Table 3**. Windows are oriented east and have a sill height 1.00 m.

2.3 Evaluation criteria

The reference used in order to evaluate the suitability of air velocity is the maximum speed allowed in classrooms (1.20 m/s). Air velocities higher than this cause occupants' dissatisfaction due to uncomfortable air currents, in addition to unwanted paper flight [17].

Thermal comfort can be evaluated through percentage of hours of occupation in thermal comfort (PHOTC) [18]. ASHRAE Standard 55 presents the procedures in order to establish PHOTC by the adaptative thermal comfort model. The indicated acceptability limit is 80%, which is recommended for typical applications [10].

Software configuration	Parameters	Adopted configuration
Geometry	Domain	Circular; height is 5 times building height and distance in the horizontal plane is 10 times the building height
	Blockage ratio	<3%
Mesh	Size function	Proximity and curvature
	Relevance center	Fine
	Smoothing	High
	Transition	Slow
	Span Angle Center	Fine (36° a 12°)
	Inflated boundary	Floor and building surfaces, maximum thickness of 0.4 and 0.2, respectively
Setup	Boundary conditions	The basis of the domain and building surfaces—wall; domain sides—inlet; top—opening
	Solver control	Interactions from 600 up to 6000

Table 4.
Specific settings in Ansys CFX.

WHO recommends a minimum natural ventilation rate equal to 10 liters per second per person in the case of non-residential indoor environments and in the COVID-19 pandemic or similar situations [19]. In relation to air humidity, WHO recommends minimum 60% and maximum 80% [20].

2.4 Simulation configuration

Ansys CFX software allows to simulate the natural ventilation provided by the different strategies. **Table 4** presents the initial parameters adopted for simulation configurations.

The adopted model of turbulence is k-ε, which optimizes the relation between the processing time and accuracy of results [21]. This is widely used in computational fluid dynamics (CFD) in order to simulate airflow patterns in turbulent conditions, mainly for low-rise buildings.

Air velocity (m/s) and pressure coefficients (Cp) on openings are the output variables. These results are exported to EnergyPlus software, where the model geometry, materials' physical characterization, and occupation schedule, among other information, are included. Regarding occupation, internal heat gains, 37 users, and a lighting power density equal to 9.90 W/m^2, that is, the density for the level maximum of energy efficiency recommended by Brazilian standards [17], are also initial simulation parameters.

For the object simulation, surface mount light is inserted; in this way, radiant fraction is 0.72, visible fraction is 0.18, and air return fraction is zero, according to Brazilian standards [22]. The simulation of natural ventilation is performed through the multizonal Airflow Network model, except for the solar chimney, which is configured as the Thermal Chimney model on EnergyPlus.

The opening factor of the frames for natural ventilation, configured in the input object Airflow Network: Multizone: Surface, depends on the percentage of opening for the ventilation of each type of window: awning has a factor of 0.9 (90% of effective area for ventilation) and maxim-ar has a factor of 0.8 (80% of effective are for ventilation) [23]. Airflow is considered at 0.95 m, which is the average respiration height of sitting children [11], and 1.5 m and 0.95 m, which are the heights of human respiration [24]. The maximum air velocity to guarantee users' comfort is 1.2 m/s [17].

With reference to windows' operation, the adopted setpoint is based on the adaptative thermal comfort [10] and on Santa Maria TMY2 [25]. When the internal temperature reaches 22°C, the windows are opened.

In order to simulate the solar chimney, the reference temperature is the same (22°C), the outlet cross section is 0.45 m^2, the inlet 0.36 m^2, the discharge coefficient is 0.8, and the chimney length is equal to 2.6 m (distance between air inlet and outlet).

Operative temperature (°C), external air temperature (°C), air changes per hour (h^{-1}), and relative air humidity (%) are the output variables obtained by EnergyPlus.

The annual school days considered add up to a total of 1600 hours according to Brazilian laws for education [26].

The analysis is based on the comparison between air velocity, occupied hours in thermal comfort, heating and cooling degree-hours, ventilation rates, air changes per hour, and relative air humidity.

3. Analysis and discussion

3.1 Indoor airflow

Table 5 illustrates the airflow pattern for the 16 strategies on plans with heights of 0.95 m and 1.5 m and for two airflow directions, azimuths 90° and 135°. Considering the minimum air velocity in order to ensure thermal comfort, 0.10 m/s [17], windows reached by a 90° wind direction are the worst solutions. There is better air distribution when wind hits the windows obliquely (135°). The higher air velocities reached are those with north–south cross-ventilation with two south-facing solar chimneys. For cross-ventilation (strategies 3, 4, 5, 6, 11, 12, 13, 14, 15, and 16), the flow pattern is also more effective for air distribution, as expected. The sill height (1.0 m) ensures that the air velocity at 0.95 does not reach values above 1.2 m/s. Strategies based on stack effect (5, 6, 13, and 14) reach the higher velocities at respiration heights.

3.2 Percentage of hours of occupation in thermal comfort (PHOTC)

Table 6 presents the results for PHOTC. Strategy 1 (single-sided ventilation, east-facing windows) is the best solution, with PHOTC equal to 80.07%. Strategies with better results are those without permanent upper openings. Strategy 4 (cross-ventilation, east–west orientation, and with permanent ventilation) is the worst result, with PHOTC equal to 70.42%. Considering the importance of permanent ventilation during the year, strategy 14 (north–south cross-ventilation and stack effect) presents the best result (73.82%). Strategies with permanent ventilation, which remain opened in the cold period of the year, between May and August, present the worst results. On the other hand, March, April, and October present the best results, as can be seen in **Table 6**, which shows the results of percentage hours in occupation in cold and heat thermal discomfort.

Strategy 1 results in the lowest cold discomfort, with 250 hours, or 15.62%, of occupied annual hours, in view of the permanent ventilation absence, whereas strategy 10 produces 404 hours, or 25.25%, of occupied annual hours in cold discomfort.

Strategy	Height 0.95 m		Height 1.5 m	
	90°	135°	90°	135°
1, 2	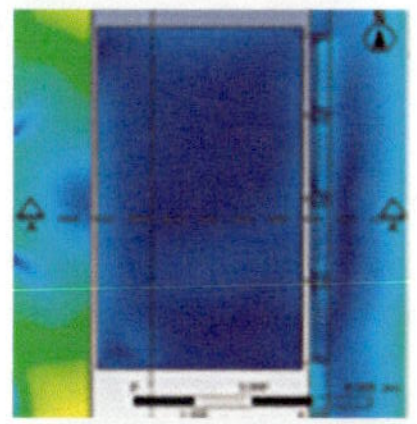	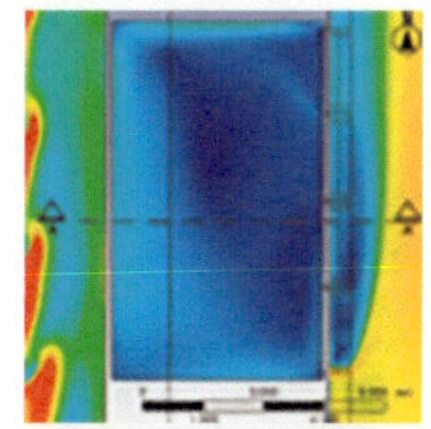	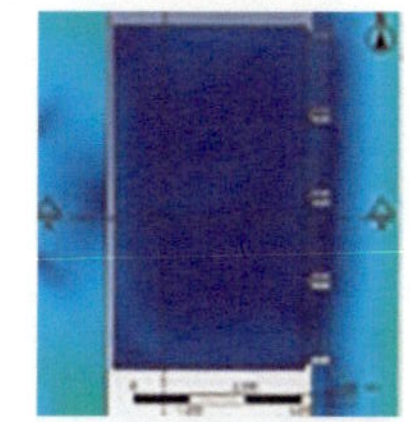	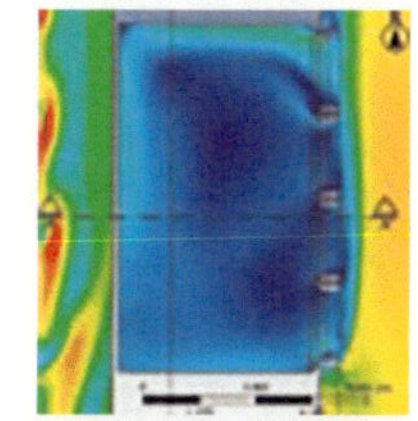
3, 4	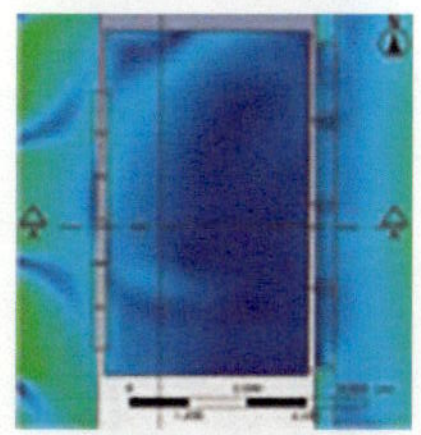	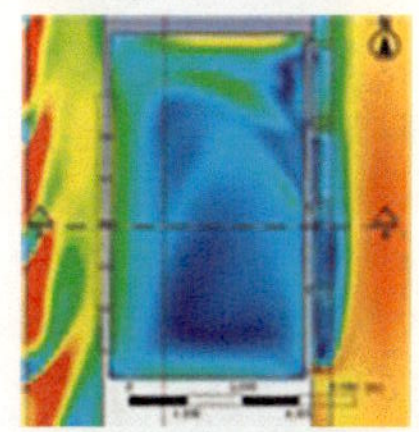	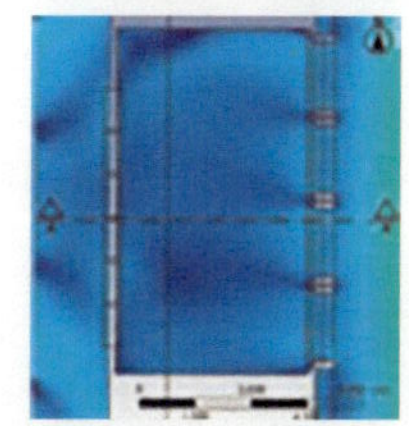	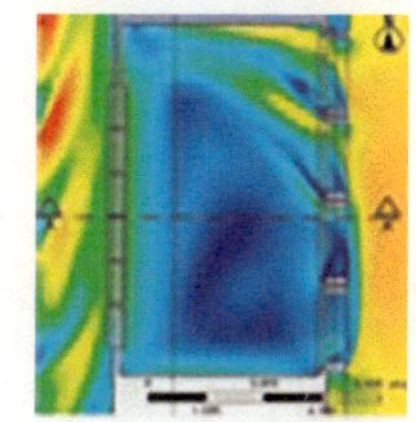
5, 6	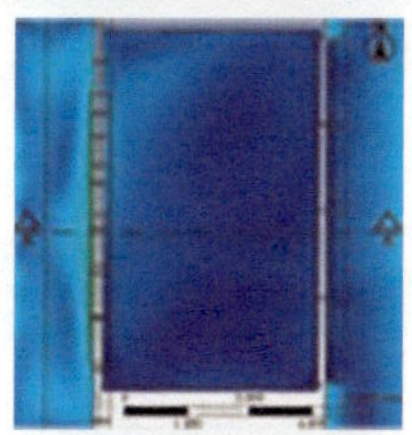	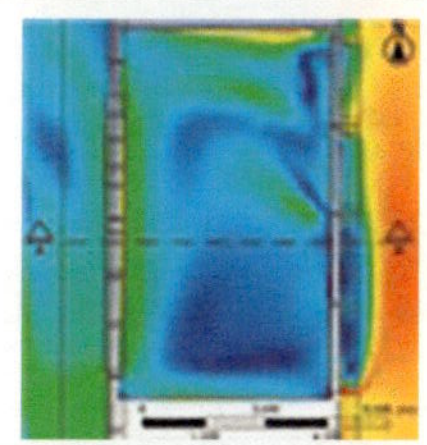	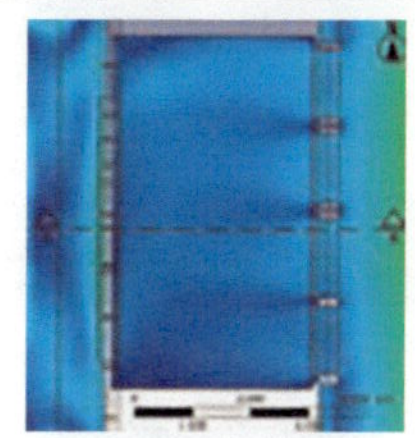	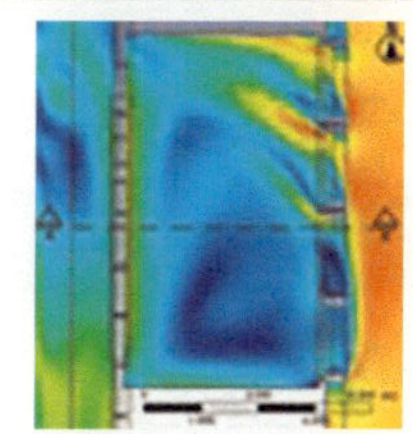
7, 8	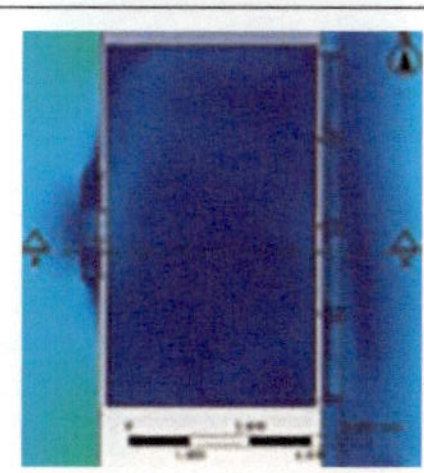	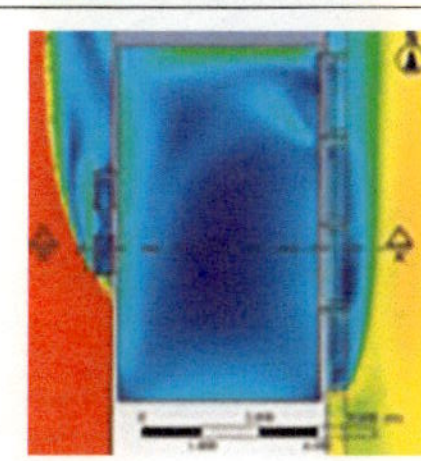	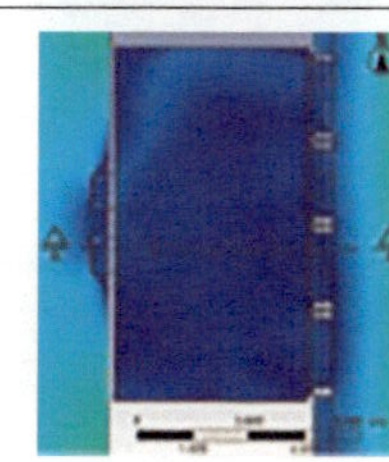	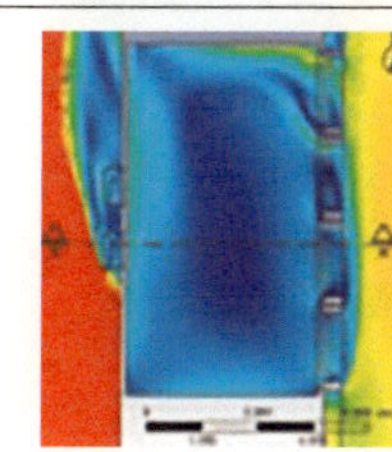

Strategy	Height 0.95 m		Height 1.5 m	
9, 10		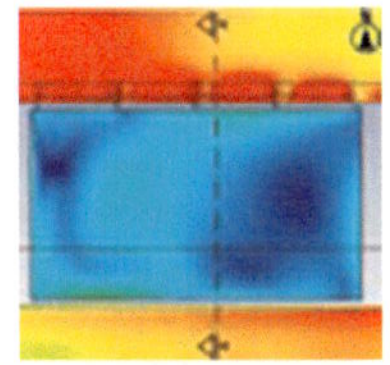	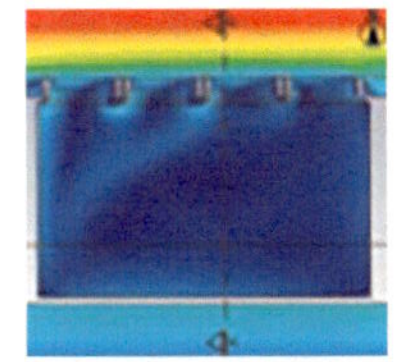	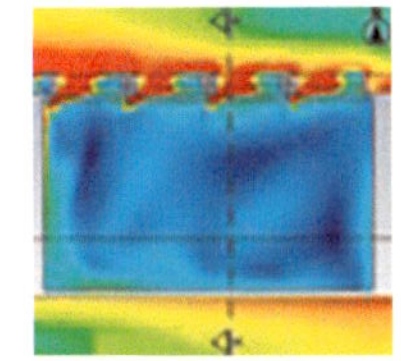
11, 12	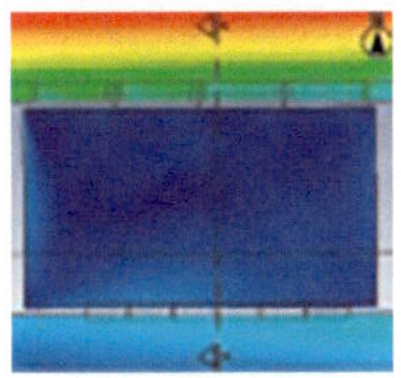	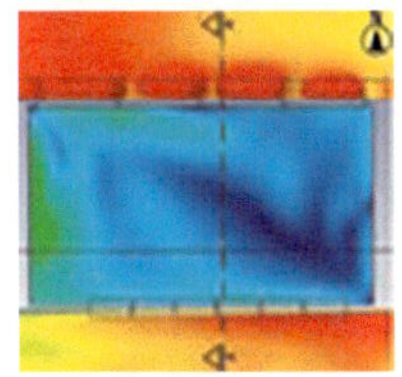	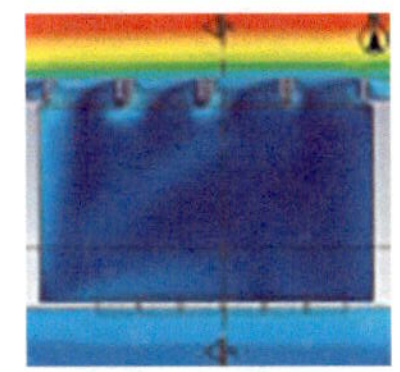	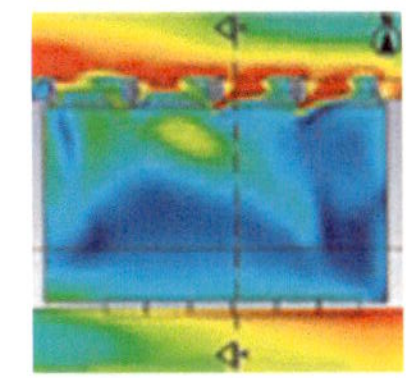
13, 14	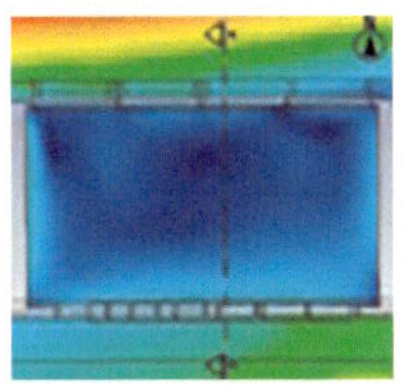	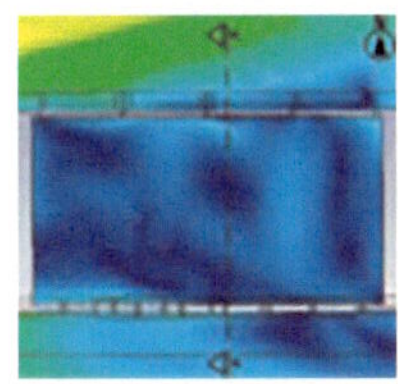	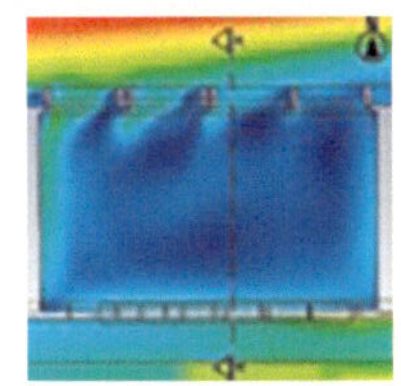	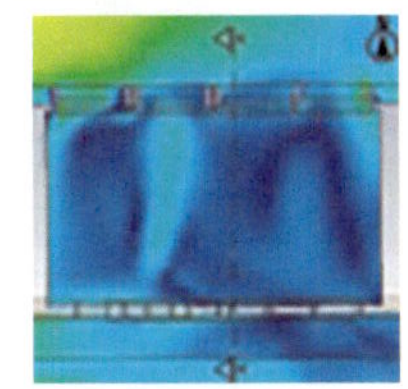
15, 16	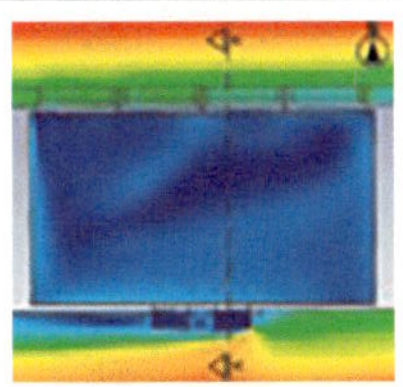	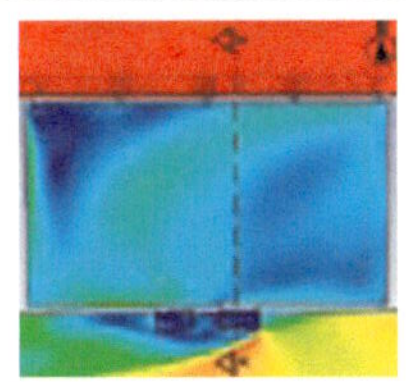	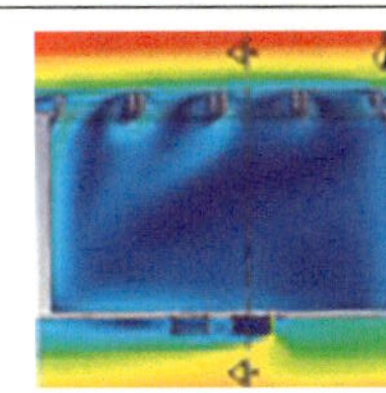	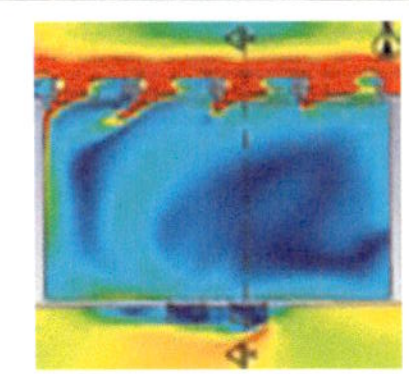
Air velocity (m/s) 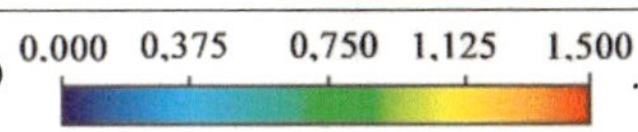				

Table 5.
Airflow pattern for the 16 strategies on plans with heights of 0.95 m and 1.5 m and for two airflow directions, 90° and 135°.

Strategy	PHOTC (%)	Strategy	PHOTC (%)
1	80.07	9	79.19
2	72.39	10	71.00
3	73.93	11	74.57
4	70.42	12	70.63
5	74.07	13	76.94
6	71.57	14	73.82
7	78.19	15	77.76
8	71.63	16	72.01

Table 6.
Annual PHOTC according to 16 strategies.

Strategy 10 differs from strategy 1 in window orientation, which indicates the importance of windows exposition to cold wind in order to ensure thermal comfort. Usually, the school period comprises the coldest months of the year; then, it is important for users' thermal comfort to consider occupied hours in cold thermal discomfort. Results show that children will be more exposed to cold conditions than heat. The maximum hour in heat discomfort is only 81 h, or about 10 days, while the maximum cold discomfort correspond to 51 days.

Considering strategies with up to 20% of hours in cold discomfort, east-facing single-sided ventilation 1 (31 days), west-facing solar chimney 7 (34 days), north-facing single-sided ventilation 9 (34 days), and north-facing solar chimney 15 (37 days) have the best performance.

Figure 3 illustrates the coldest and hottest weeks during the school year. In the coldest week, for all strategies, users will be in cold discomfort practically all the time. Strategies 1 (single-sided ventilation, east orientation), 7 (west-facing solar chimneys), and 9 (single-sided ventilation, north orientation) present the best results, with 15, 14, and 14 h in thermal comfort, respectively, during this week. In this case, window closure is recommended; teachers must have close windows according to users' cold sensation and the perception of level of ventilation required. In contrast, for the hottest week of the school year, most of the occupied hours are thermally comfortable. The best strategies are 13 and 14 (north–south oriented with stack effect), with 35 and 36 hours in thermal comfort, respectively.

3.3 Heating and cooling degree hours

In order to analyze annual heating degree hours (HDHr) and cooling degree hours (CDHr), the reference 80% of acceptability is considered [10]. **Figure 4** presents the results according to heating or cooling and different analyzed strategies.

Strategy 1 (single-sided ventilation, east orientation) presents the lowest HDHr (408.37 degree-hours) and for annual heating and cooling (468.67 degrees-hours). Strategy 13 (north–south cross-ventilation, stack effect) presents the lowest CDHr (42 degree-hours). Strategies 7 (west-facing solar chimneys), 9 (north single-sided ventilation), and 15 (south-facing solar chimneys) also present an acceptable performance, with up to 500 degree-hours for heating. Window orientation in relation to wind direction is crucial for the performance of strategies. Although the wind direction of 135° is better for air distribution and level of

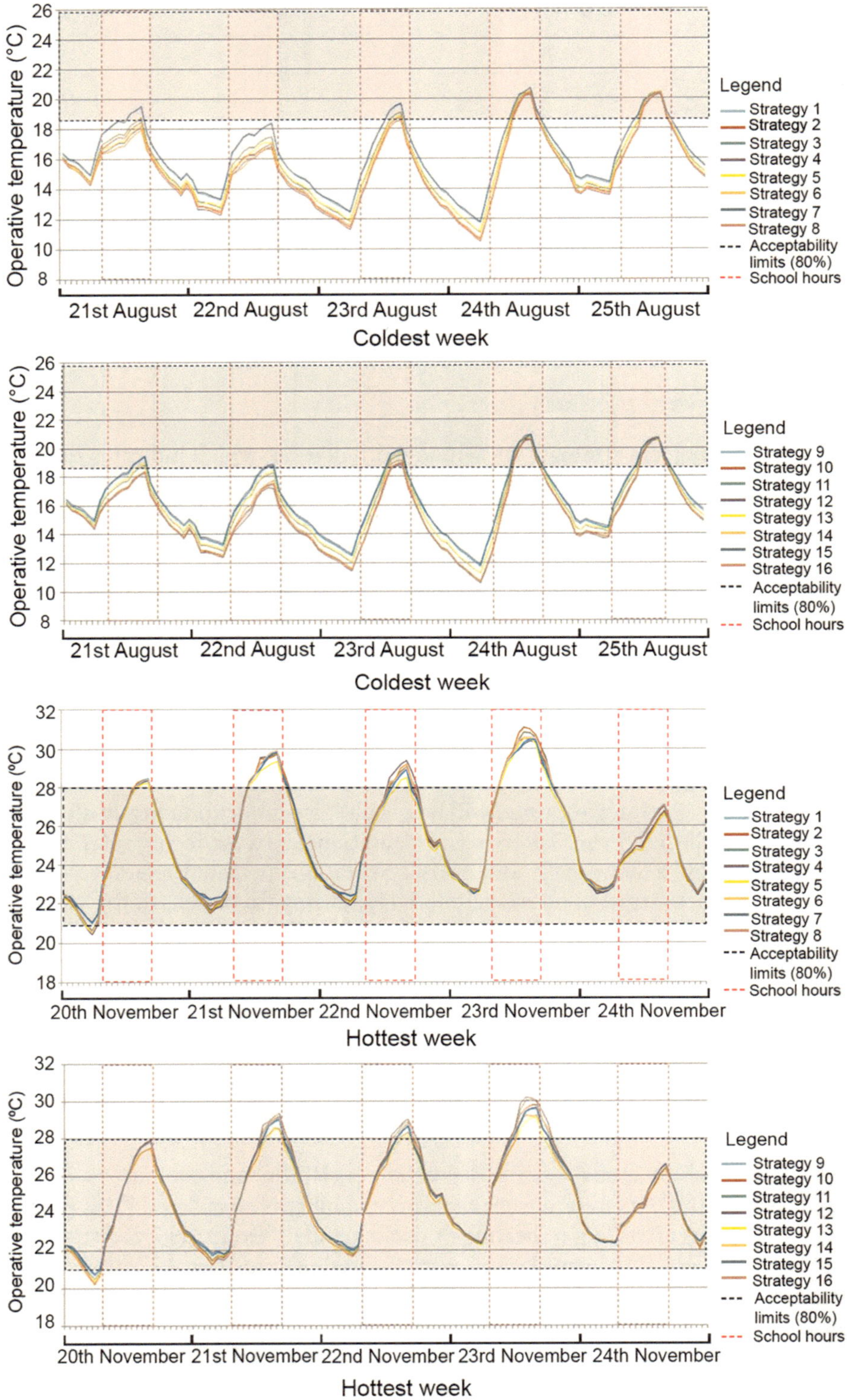

Figure 3.
Operative temperature in the coldest and hottest weeks of the school year.

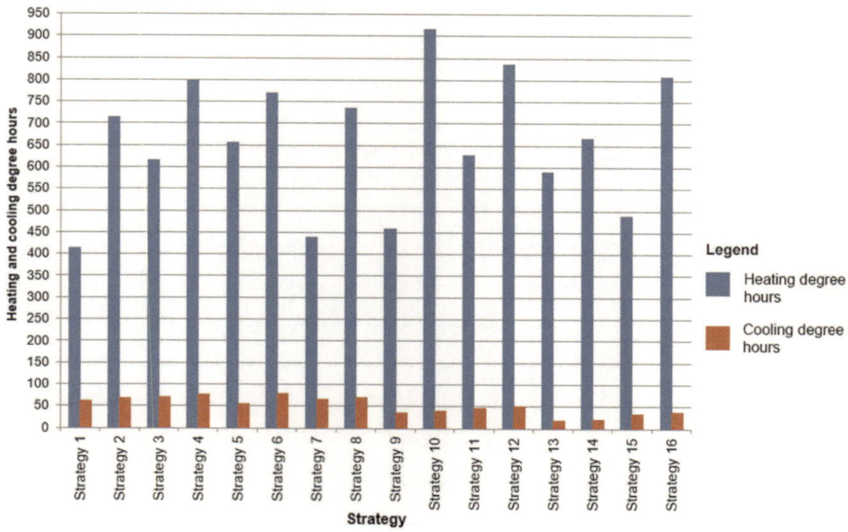

Figure 4.
Heating and cooling degree hours for each strategy.

ventilation required, it causes high cold discomfort due to the school year being predominantly in the coldest months of the year. This result demonstrates the difficulty in solving effective ventilation expressed by ventilation rates and airflow patterns and the users' thermal comfort.

3.4 Ventilation rate and air changes per hour

Considering that 10 l/s/p is the recommended rate, the ventilation rates reached by different strategies demonstrate the insufficiency of single-sided ventilation. Strategies 1, 2, 9, and 10 present rates lower than 3 l/s/p. On the other hand, strategies 3–6 and 11–14, which are cross-ventilated solutions, present high rates, which exceed 38 l/s/p. Strategies based on solar chimneys (7, 8, 15, and 16) reach 13 l/s/p, presenting the best performance.

In relation to air changes per hour, **Table 7** presents results for the 16 strategies considering the annual average. The reference is 10.16 changes per hour [27]. Most of the strategies (69%) present values below that indicated. Strategies that reached the recommended value are those which are based on cross-ventilation and stack effect, east–west oriented, except for 12, which is north–south cross-ventilation with permanent ventilation.

Considering these results and the percentage of hours in cold discomfort (**Table 8**), the best results (the lowest percentages) correspond to 1, 7, 9, and 15, as already commented. However, these strategies present the lowest changes per hour, compromising indoor air quality.

The second set of strategies with lower cold discomfort is represented by 3 (21.31% or 43 days), 5 (21.75% or 44 days), 11 (21.37 or 43 days), and 13 (20.44% or 41 days). For these, air change per hour is equal to 10.56, 12.84, 9.44, and 8.53 c.p.h., respectively. When considering the heating degree-hours, these strategies also have a medium performance, presenting 600 degree-hours, 650 degree-hours, 625 degree-hours, and 600 degree-hours, respectively.

Strategy	Annual average of air changes per hour
1	7.65
2	8.19
3	10.56
4	11.00
5	12.84
6	13.95
7	7.68
8	7.98
9	7.65
10	8.16
11	9.44
12	10.29
13	8.53
14	8.82
15	7.81
16	8.34

Table 7.
Annual hourly air changes for each case.

Strategy	Percentage of hours in cold discomfort (%)	Percentage of hours in heat discomfort (%)	Total percentage of hours in thermal discomfort (%)
1	15.62	4.31	19.93
2	22.88	4.81	27.61
3	21.31	4.81	26.07
4	24.64	4.94	29.58
5	21.75	4.18	25.93
6	23.37	5.06	28.43
7	16.87	4.94	21.81
8	23.37	5.00	28.37
9	17.19	3.62	20.81
10	25.25	3.75	29.00
11	21.37	4.06	25.43
12	25.06	4.31	29.37
13	20.44	2.62	23.06
14	23.37	2.81	26.18
15	18.62	3.62	22.24
16	24.37	3.62	27.99

Table 8.
Percentages of hours occupied in cold and heat discomfort.

These strategies are based on east–west cross-ventilation, 90° wind direction (3); west-facing stack effect, 90° wind direction (5); north–south cross-ventilation, 90° wind direction (11); and south-facing stack effect, 90° wind direction (13).

The need to achieve thermal comfort in cold periods and the level of ventilation required leads to the adoption of intermediate performance; that is, the strategies will not present the best performance for all criteria but will ensure values as close as possible to all recommended criteria. For extreme situations, a mechanical system for air renewal will be necessary.

3.5 Relative humidity

Considering the range of relative humidity recommended by WHO [10], 60–80%, for monthly average, all strategies reach the criteria. Strategies with cross-ventilation, stack effect or solar chimneys, and permanent ventilation (located near the ceiling) present lower values for relative humidity, as expected since hot humid air (e.g., generated by users' respiration and sweat) accumulated on the top of classrooms is more easily removed by cross-ventilation than ones with single-sided ventilation or without permanent ventilation.

4. Final considerations

Regarding cities in subtropical humid climate and developing countries, represented by the Brazilian southern city Santa Maria, results demonstrated that the demand for the required ventilation may compromise users' thermal comfort mainly in the cold period of school year. In this case, natural ventilation systems must consider the balance between thermal comfort and air exchange efficiency.

Strategies based on cross-ventilation or stack effect and permanent ventilation present lower percentage of hours in thermal comfort for the coldest period of the school year but present air changes in accordance with values recommended by WHO in a pandemic context. On the other hand, satisfactory thermal comfort levels and air change may also be reached with solutions based on cross-ventilation and stack effect when the window orientation relative to wind is 90°; that is not the best orientation for airflow pattern.

In addition, single-sided ventilation is not recommended since natural ventilation rates provided by it does not reach even half criteria (10 l/s/p).

This study presents a contribution to the understanding of the interrelationships between children's thermal comfort and the need for high ventilation rates in classrooms in the face of a pandemic context. In developing countries, where the use of mechanical or air conditioning systems is limited by economic constraints, passive technologies are important in order to achieve healthy and comfortable environments.

Author details

Paula Scherer, Daiana de Oliveira Fauro and Giane de Campos Grigoletti*
Universidade Federal de Santa Maria, Santa Maria, Brazil

*Address all correspondence to: giane.c.grigoletti@ufsm.br

References

[1] Jones NR, Qureshi ZU, Temple RJ, Larwood JPJ, Greenhalgh T. Two metres or one: what is the evidence for physical distancing in COVID-19?. British Medical Journal. 2020;**370**:m3223. DOI: 10.1136/bmj.m3223

[2] Franceschini PB, Liguori IN, Neves LO. Avaliação da qualidade do ar interior durante a pandemia de COVID-19 em salas de aula naturalmente ventiladas. In: Encontro Nacional de Conforto no Ambiente Construído, 26; Encontro Latino-Americano de Conforto no Ambiente Construído, 22. Palmas, 2021. Anais [...] Porto Alegre: ANTAC; 2021

[3] Hou D, Katal A, Wang LL. Bayesian calibration of using CO_2 sensors to assess ventilation conditions and associated COVID-19 airborne aerosol transmission risk in schools. Medrxiv Preprint. fev. 2021. pp. 1-42

[4] Ding E, Zhang D, Bluyssnen PMO. Avaliação da qualidade do ar interior durante a pandemia de COVID-19 em salas de aula naturalmente ventiladas. In: Healthy Buildings Europe 2021 Online Conference. Trondheim, 2021. Proceedings [...] Trondheim: ISIAC; 2021

[5] Barrett P, Davies F, Zhang Y, Barrett L. The impact of classroom design on pupil's learning: Final results of a holistic, multi-level analysis. Building and Environment. 2015;**89**:118-133

[6] Kowaltowski DCCK. Arquitetura escolar: o projeto do ambiente de ensino. São Paulo: Oficina de Textos; 2011

[7] Rackes A, Fonseca RW, Beck EO, Scalco VA, Palladini GD, Lamberts R. Avaliação do potencial de conforto térmico em escolas naturalmente ventiladas. In: Proceedings of the National Meeting on Comfort in the Built Environment (ENCAC '13); 15–16–17 October 2015; Campinas. Porto Alegre: ANTAC; 2016. pp. 1-10

[8] Maciel LF. Brise-chaminé-solar: avaliação experimental e por simulação CFD/Energyplus de um dispositivo de ventilação [thesis]. Viçosa: Federal University of Viçosa; 2016

[9] Pagel EC, Gouveia GLO, Martins RS, Cruz MVG. Ventilação natural e desempenho térmico sob diferentes configurações de aberturas em uma sala de aula. Ambiente Construído. 2022;**22**: 133-157. DOI: 10.1590/s1678-86212022000300612

[10] ASHRAE. ANSI/ASHRAE Standard 55–2017 Thermal Environmental Conditions for Human Occupancy. Atlanta: ASHRAE (American Society of Heating, Refrigerating and Air Conditioning Engineers); 2017

[11] Dreyfuss H. As medidas do homem e da mulher: fatores humanos em design. Porto Alegre: Bookman; 2005

[12] INMET. Normais Climatológicas do Brasil. 2021. Available from: https://portal.inmet.gov.br/normais. [Accessed: 05 February 2022]

[13] Flores MG. Geração da base climática de Santa Maria-RS—para análise de desempenho térmico e eficiência energética de edificações [thesis]. Santa Maria: Federal University of Santa Maria; 2016

[14] Weber FS, Melo AP, Marinoski D, Lamberts R. Elaboração de uma biblioteca de componentes construtivos brasileiros para o uso no programa

EnergyPlus. 2017. Available from: https://labeee.ufsc.br/node/714. [Accessed: 05 December 2022]

[15] ABNT. NBR 15220-2:2005 Métodos de cálculo da transmitância térmica, da capacidade térmica, do atraso térmico e do fator solar de elementos e componentes de edificações. Rio de Janeiro: ABNT (Brazilian Association of Technical Standards); 2005

[16] ABNT. NBR 15220-3:2005 Desempenho térmico de edificações Parte 3: Zoneamento bioclimático brasileiro e diretrizes construtivas para habitações unifamiliares de interesse social. Rio de Janeiro: ABNT (Brazilian Association of Technical Standards); 2005

[17] Bayoumi M. Improving indoor air quality in classrooms via wind-induced natural ventilation. Modelling and Simulation in Engineering. 2021;**2021**: 1-14. DOI: 10.1155/2021/6668031

[18] BRASIL. Ordinance No. 42, of February 24, 2021. Approves the Inmetro Normative Instruction for the Energy Efficiency Classification of Commercial, Service and Public Buildings (INI-C); 2021

[19] WHO. Roadmap to Improve and Ensure Good Indoor Ventilation in the Context of COVID-19. World Health Organization. Geneva: WHO (World Health Organization); 2021. p. 25

[20] OMS. Actas Oficiales de La OMS [Internet]. 2012. Available from: https://apps.who.int/gb/or/s/. [Accessed: 05 December 2022]

[21] Bevilaqua CP, Bressianini B, Azuma MH, Lukiantchuk MA. Análise da ventilação natural. PARC Pesquisa em Arquitetura e Construção. 2019;**10**: e0190221-e01902213. DOI: 10.20396/parc.v10i0.8652835

[22] ABNT. NBR 16401-2:2008 Instalações de ar-condicionado—Sistemas centrais e unitários. Rio de Janeiro: ABNT (Brazilian Association of Technical Standards); 2008. p. 7

[23] PBE EDIFICA. Manual para a aplicação do RTQ-R. Rio de Janeiro: Eletrobras/Procel (National Electric Energy Conservation Program); 2012. p. 202

[24] Morais JMS. Ventilação natural em edifícios multifamiliares do programa "Minha Casa Minha Vida". Campinas, 2013.211 f. Tese (Doutorado em Arquitetura, Tecnologia e Cidade). Campinas: Faculdade de Engenharia Civil, Universidade Estadual de Campinas; 2013

[25] Buligon LB. Comportamento higrotérmico e energético de painéis de vedação vertical externa em madeira para a zona bioclimática 2 [thesis]. Santa Maria: Federal University of Santa Maria; 2021

[26] BRASIL. Lei n. 9.394, de 20 de dezembro de 1996. Estabelece as diretrizes e bases da educação nacional. 1996. Disponível em: http://www.planalto.gov.br/ccivil_03/leis/L9394compilado.htm. [Acessoem: 25 jan. 2021]

[27] OPAS. 2021. Disponível em: https://iris.paho.org/handle/10665.2/53938. [Acessoem: 9 out. 2021]